Repair of Concrete Structures to EN 1504

Repair of Concrete Structures to EN 1504

A guide for renovation of concrete structures – repair materials and systems according to the EN 1504 series

Dansk Standard
(Danish Standards Association)

CRC Press
Taylor & Francis Group
Boca Raton London New York

CRC Press is an imprint of the
Taylor & Francis Group, an **informa** business

First published by Butterworth-Heinemann

First published 2004

This edition published 2011 by Routledge
2 Park Square, Milton Park, Abingdon, Oxon OX14 4RN
711 Third Avenue, New York, NY 10017, USA

Routledge is an imprint of the Taylor & Francis Group, an informa business

British Library Cataloguing in Publication Data
A catalogue record for this book is available from the British Library

Library of Congress Cataloguing in Publication Data
A catalogue record for this book is available from the Library of Congress

ISBN: 978-0-7506-6222-2

Contents

7 Conditional repair materials and systems 45

8 Maintenance after rehabilitation 47

9 Health, safety and environment 48

Preface

This handbook is a guide for the EN 1504-series: 'Products and systems for protection and repair of concrete structures. Definitions, requirements, quality control and evaluation of conformity'. The EN 1504 series is a number of standards on products and systems for the protection and repair of concrete structures, including:

EN 1504-1	Part 1: Definitions
EN 1504-2	Part 2: Surface protection systems
EN 1504-3	Part 3: Structural and non-structural repair
EN 1504-4	Part 4: Structural bonding
EN 1504-5	Part 5: Concrete injection
EN 1504-6	Part 6: Anchoring products
EN 1504-7	Part 7: Reinforcement corrosion protection: coatings for reinforcement
EN 1504-8	Part 8: Quality control and evaluation of conformity
ENV 1504-9	Part 9: General principles for the use of repair materials and systems
EN 1504-10	Part 10: Site application of products and systems, and quality control of the works

On publication of the handbook the following parts in the above-mentioned EN 1504 series were available as drafts: Parts 2, 3, 4, 5, 6, 7, 8 and 10.

The standards in the above series are brief, as standards should be, and only contain main principles, not actual 'textbook material'. The above standards contain guiding annexes. It is the opinion of the Danish Standards Institution, however, that a supplementary, guiding text will be necessary for the standards.

This guide for the EN 1504 standards addresses the principles and methods that form the basis of the choice of repair materials and systems for rehabilitation of damaged concrete structures.

Annex C of the guide includes a complete overview of the CEN and ISO test standards which, according to the EN 1504 series, are applicable for documentation of conformity with the requirements.

It is pointed out that certain test methods referenced in ENV 1504-9 have been changed.

Annex E of the guide is related to the above-mentioned standards and describes specific aspects that:

- manufacturer/supplier
- designer/builder
- contractor/supervisor

should take into consideration in their choice of repair materials and systems. The purpose of this description is to provide an overview of the requirements made and to explain how conformity with these requirements can be documented. Annex A on determination of the characteristic value of tests made is related to this description.

It is the opinion of the Danish Standards Association, however, that these standards are so close to their final version that a Danish guide will be justified, so that the use of the rehabilitation standards may be incorporated in rehabilitation construction work before the official approval some time in 2004.

This situation means that some of these drafts from 2002 are rather incomplete. Therefore, they are likely to be subject to change with regard to details before they can be adopted. This applies to, for example, standard EN 1504-7 on 'Reinforcement corrosion protection'.

This guide is prepared by the task force, S-328/U-06, Repair and protection of concrete structures. The object of the task force is to follow the development and comment on drafts from CEN on proposed new test methods and standards within the field of rehabilitation of concrete structures.

Project participants

Task force:
Arne Damgaard Jensen Teknologisk Institut, Betoncentret
Erik Stoklund Larsen, chairman   COWI A/S
Ervin Poulsen, technical secretary   Ervin Poulsen aps
Jens M. Frederiksen   NIRAS/AEClaboratoriet A/S
Jørgen Schou   Fosroc A/S
Kaj Lund   Skanska Danmark A/S
Lotte Torp   MT Højgaard A/S
Niels Damsager Hansen   COWI A/S
Ole Viggo Andersen   NIRAS/AEClaboratoriet A/S
Tommy B. Jacobsen   Teknologisk Institut, Betoncentret

This publication is a guide for the application of EN 1504 standards and other standards for rehabilitation and maintenance of concrete structures. Not all of these standards have been finally adopted. It is the opinion of the Danish Standards Association, however, that these standards are so close to their final version that a Danish guide will be justified, so that the use of the rehabilitation standards may be incorporated in rehabilitation of construction work before the official approval some time in 2004.

The guide will always compare the latest version of EN 1504 standards to the other standards on rehabilitation and maintenance. The latest issue of the English version will always be valid.

Introduction

Protection, repair and reinforcement (i.e. rehabilitation) of concrete structures requires relatively complicated design, dimensioning and control. The series of CEN standards under EN 1504 defines the principles of rehabilitation of concrete structures, which are damaged. Furthermore, these standards specify guidelines for the choice of repair materials and systems that are appropriate for rehabilitation and maintenance of concrete structures. These standards describe the main points of rehabilitation of a damaged concrete structure:

- Assessment of the registered state of a concrete structure
- Determination of the courses of damage
- Determination of the objective of the rehabilitation of a damaged concrete structure
- Choice of relevant principles for rehabilitation of a damaged concrete structure
- Choice of methods for rehabilitation of a damaged concrete structure
- Definition of the properties for repair materials and systems for rehabilitation of a damaged concrete structure or its members
- Specification of requirements for the maintenance that should always follow rehabilitation of a damaged concrete structure or its members.

It is significant to note that the EN 1504 standards do not exclude other methods than those mentioned in ENV 1504-9. However, application of such methods is limited to situations in which their application is justified. However, documentation of the properties of the considered repair materials and systems and their characteristics is mandatory.

1

Scope

The EN 1504 standards specify conditions to be taken into account in the design, construction and control of rehabilitation of damaged concrete structures (reinforced and non-reinforced) using repair materials and systems as specified in the EN 1504 standards and other relevant EN standards or recognized European technical regulations. The EN 1504 standards contain the following:

1.1 Registration of state

The EN 1504 standards specify the necessity of carrying out inspection, testing and evaluation before, during and after rehabilitation of a damaged concrete structure or its members.

1.2 The environment

The EN 1504 standards specify how to rehabilitate structures with defects and impairments caused by environmental action, as defined in EN 206-1, or other chemically aggressive occurrences.

The resistance of concrete and concrete structures in different exposure classes is defined in EN 206-1. The EN 1504 standards deal with such repair of materials and systems that are necessary for rehabilitation of damage caused by chemical attack, as defined in EN 206-1, or other chemical attack. The general principles in ENV 1504-9 may be applicable for choosing repair materials and systems for rehabilitation of damage caused by more aggressive chemical attack than those described in EN 206-1, but further measures will be assumed to be necessary.

Biological attack: Certain bacterial species may have decomposing effect on concrete by liberating corroding substances, such as nitric acid or sulphuric acid. This is a chemical attack on the concrete. Sulphur bacteria are widespread in nature. In addition to corrosion on the concrete, they may also result in sulphate attack.

Concrete cover: The minimum values of concrete cover on reinforcement, as specified (required) in EN 206-1, can protect the concrete reinforcement against corrosion in the normal, natural exposure classes (as defined in EN 206-1), inclusive of marine environments and application of chloride-containing de-icing salts. This does not necessarily apply to concrete structures which do not fulfil the requirements of EN 206-1, e.g. with constituents which are not conditional (according to EN 206-1) or which have inadequate concrete cover (with regard to thickness and/or proportioning).

1.3 Mechanical action

The EN 1504 standards specify rehabilitation of damage and defects caused by mechanical action, such as incongruous settlement, excessive loads (inclusive of earthquakes and impact action), organic attack, inadequate design and construction, or application of inappropriate construction materials.

Organic attack: Concrete may be subject to attack from piddocks and certain bristle worms, which can make up to 100 mm deep holes with a diameter of 4–12 mm, particularly in aggregates of limestone. This situation, however, is insignificant in the Nordic countries (due to requirements for concrete aggregates), but in southern Europe and in tropical countries it is not an insignificant problem.

1.4 Alkali reaction

The EN 1504 standards also specify rehabilitation with the purpose of reducing the effects of alkali reaction activities.

1.5 Reinforcement

The EN 1504 standards specify reinforcement of damaged concrete structures by:

* Supplementing embedded concrete or placement of external reinforcement (e.g. cables, flat rolled steel and carbon fibre strips)
* Filling (injection) of cracks, joints and interstices between concrete elements to create the necessary structural integrity
* Reinforcement of damaged concrete structures by increased concrete dimensions.

1.6 Preventive maintenance

The EN 1504 standards specify waterproofing as a necessary part of rehabilitation by preventive maintenance of damaged concrete structures. Further, rehabilitation of direct-laid concrete is specified. Methods for preventive maintenance also include:

* Treatment of cracks
* Recreation of the passivity of the reinforcement
* Reduction of the corrosion rate of the reinforcement by limiting the moisture content of the concrete. It shall be noted, however, that reducing the moisture content of the concrete due to driving rain by structural measures (coating, overhang, etc.) is not efficient if the concrete can absorb moisture in another way (e.g. capillary absorption)
* Reduction of the corrosion rate of the reinforcement by electrochemical methods. The electrochemical methods (re-alkalization, chloride extraction and cathodic protection) are included due to the ability of these methods to reduce the corrosion rate of the reinforcement in cases where it is impractical or uneconomic to remove carbonated or chloride infected concrete and replace it by new concrete or mortar
* Control of the corrosion rate of the reinforcement by surface protection.

1.7   Execution

Execution of the methods in ENV 1504-9 for rehabilitation of concrete structures is described in more detail in EN 1504-10.

1.8   Exceptions

The EN 1504 standards do not, however, include the requirements for rehabilitation of concrete structures dealing with:

- Damage due to fire
- Repair materials and systems to be used for purposes other than rehabilitation of damaged concrete structures, e.g. to improve the aesthetic appearance or change the primary use of a concrete structure
- Defects and damage in post-tensioned concrete structures.

It should be noted, however, that the general principles in ENV 1504-9 are applicable in such cases. Furthermore, it should be noted that site application is solely specified in EN 1504-10 on 'Site application of products and systems and quality control of the works'. EN 1504-10 includes:

- Preparation of concrete and reinforcement before application of repair materials and systems for rehabilitation
- Minimum requirements, for climatic conditions for storing repair materials and systems
- Control of the quality of rehabilitation works.

2

Related standards

The European standards for rehabilitation of damaged concrete structures are based on extensive dated and undated references to other publications. These references are given in the text. An overview of test standards is given in Annex C. The standards for rehabilitation and maintenance of concrete structures include the EN 1504 standards and related standards.

ENV 1992-1-1: Design of concrete structures. General rules and rules for buildings (Eurocode 2)

EN 206-1: Concrete – Execution, production and control

EN 12696-1: Cathodic protection of steel in concrete – Part 1: Atmospherically exposed concrete

EN 14038-1: Electrochemical re-alkalization and chloride extraction treatment for reinforced concrete – Part 1: Re-alkalization

EN 14487-1: Sprayed concrete – Part 1: Definitions, specifications and conformity

EN 1504-series: Products and systems for the protection and repair of concrete structures – Definitions – Requirements – Quality control and evaluation of conformity (includes the following standards):
- EN 1504-1   Part 1: General scope and definitions
- EN 1504-2   Part 2: Surface protection systems
- EN 1504-3   Part 3: Structural and non-structural repair
- EN 1504-4   Part 4: Structural bonding
- EN 1504-5   Part 5: Concrete injection
- EN 1504-6   Part 6: Anchoring products
- EN 1504-7   Part 7: Reinforcement corrosion prevention
- EN 1504-8   Part 8: Quality control and evaluation of conformity
- ENV 1504-9   Part 9: General principles for the use of products and systems
- EN 1504-10   Part 10: Site application of products and systems and quality control of the works

The guide is based on the standards available in 2002.

3

Definitions and explanation of terms

EN 1504-1 on 'Definitions' specifies the following definitions and terms, which cannot be considered to be general, but are specific for the EN 1504 standards. In the other standards, EN 1504-2 to EN 1504-10, supplementary and more extensive definitions are given. These and the supplementary definitions are also included in the following list:

Absorption
Homogeneous intake of a substance into another.

Acid
An aqueous solution is considered to be acid when the pH value is below 7 at 22 °C.

Acrylic resin
Bonding agent in some paints with or without content of solvents.

Acrylics
Name for bonding agents in some paints.

Active surface protection of concrete
Surface protection (e.g. impregnation and paint) which contains active substances. When applied to a concrete surface, it reacts with the set cement paste so that the large pores on the concrete surface are filled with a crystallized product.

Active surface protection of reinforcement
Surface protection (e.g. paint) which either contains electrochemically active substances and acts as a corrosion inhibitor when applied to reinforcement, or which acts as a sacrificial anode, and thus gives cathodic protection to the reinforcement.

Additive
Pulverulent inorganic material that can be added to repair products to improve certain properties or to obtain special properties. There are two types of additives, namely:

- Type I: almost inactive additives
- Type II: puzzolane or latent hydraulic additive.

Additives for hydraulic binders
Products added to a hydraulic binder to obtain special properties and which are not covered by the term 'additives'.

Additives for reactive polymers
Other products than additives to add specific properties to repair products, e.g.:

- plasticizers
- additives to give better deformability
- accelerating additives
- retarding additives
- additives to regulate rheology
- pigments
- filler.

Adhesion by slant shear strength
Compressive (or tensile) testing of two cube specimens of steel or concrete which are glued together by scarf jointing. By measuring the cube strength at different inclinations of the scarf joint, the failure criterion of the glue can be determined provided that the glue cracks (see EN 12188).

Admixture for concrete
Material added to concrete in the mixing process (shall not exceed 5% of the cement quantity) to modify the properties of the fresh or hardening concrete.

Adsorption
Intake of a substance to the surface of another.

Affinity
A substance in concrete is said to have affinity to another material in the concrete if the substance is migrating towards the material.

Alkaline
See Basic.

Alkyd (resin)
Bonding agent formed by condensation of glycerol with (oil) acids. Sets with the oxygen in the air.

Alumina cement
A type of cement where the composition differs from Portland cement. Concrete and mortar with alumina cement as a bonding agent may lose a significant part of their strength at special temperature and moisture conditions. Therefore, concrete and mortar for which alumina cement is used wholly or partly as a bonding agent are not applicable to structural repair.

Amorphous
A substance is said to be amorphous (as opposed to crystalline) if composed of molecules which are spaced at random, almost as in a liquid.

Analysis of state
Determination of causes of damage based on registration of state supplemented with a thorough concrete analysis (site testing and laboratory tests of samples drawn).

Anchorage product
Glue where the setting is based on a reactive polymer binder, e.g. polyester. It is applied as a liquid or paste with the purpose of filling and setting so that bolts, reinforcing bars and anchors can be fastened with proper durability and strength in concrete.

Anode
The positive pole of an electric battery. Existence of corrosion in an aqueous corrosion cell (consisting of anode and cathode). An external component, e.g. a steel grid to which positive voltage is applied, e.g. by electrochemical methods.

Ballotini
Transparent glass balls with a diameter between 125 and 250 μm used for determination of the surface dryness of a paint according to EN ISO 1517.

Basic
An aqueous solution is said to be basic if the pH value increases 7 at 22 °C.

Batch
A quantity of material produced in a single operation. In case of continuous production, a batch is the quantity of material which is produced over a given (short) period of time and is assumed to have a uniform composition.

Batch quantity
See Batch.

Blast cleaning
Cleaning of a concrete surface by blast cleaning (e.g. sandblasting or high-pressure washing) so that a layer of maximum 2 mm is removed.

Bond improvement
Part of a product or a system for repair with the purpose of improving bonding between the substrate and the applied mortar or poured concrete to achieve permanent bonding which is not affected by moisture or alkalis.

Bonding
The adhesion created by applying a product or system to substrate.

Capillary absorption
The ability of a product or a system to absorb water without application of hydrostatic pressure. It is effected by capillarity where the water rises in the thinnest capillary tubes.

CAPO test
Test method for site determination of the compressive strength of concrete by pull-out tests (cut and pull out test).

Carbonation buffer
Ability in concrete or mortar to reduce carbonation.

Cathode
(1) The negative pole of an electric battery. (2) Protected part of reinforcement and metal in a building component exposed to electrochemical method. (3) Antipole to the anode in a corrosion cell.

Cement mortar and (cement) concrete
Cement based material with largest aggregate size up to 4 mm (mortar) or largest aggregate size above 4 mm (concrete). The aggregate should be graduated. The bonding agent should be hydraulic, and cement mortar (and concrete) may contain admixtures and (mineral) additions. When water is added, mortar and concrete will set by a hydraulic (under water) reaction.

Cement mortar for injection
Mix of cement, aggregates of, for example, silica dust or fine-grained sand, water and additives.

Cement paste for injection
Mix of cement, water and additives.

Characteristics
Parameters that under well-defined conditions can describe, or be applied as measures for a material, a building component or a process, etc. Characteristics are not material properties (see Properties). Thus, the characteristics of layer thickness of a film of paint are not a property of the paint.

Compatibility
Two or more substances are considered to be compatible if they are not included in a reaction which is harmful for the concrete. Porous flint is not compatible with high-alkali cement.

Compliance criteria
Requirements for, for example, measurement of properties during testing.

Concrete skin
A form of offsetting on a concrete surface. Concrete skin may be formed due to bleeding, imperfect hydratization or defective protection against desiccation.

Contraction joint
A concrete coat without joints will get irregular cracks for each 4–8 m due to moisture (shrinkage) and temperature changes. A contraction joint is made with an interstice (possibly sawn) and may be through-going. The joint is filled with joint filler and joint pad or pre-compressed, impregnated joint strips.

Control scheme
A programme to ensure accomplishment of an activity according to planned intentions.

Copolymer
Polymer made from two or more different monomers.

Corrosion protection of reinforcement
Corrosion protection (e.g. paint) applied to the reinforcement to insulate it from the pore water in the ambient mortar and concrete.

Crack bridging
A paint film is considered to bridge a crack if it is able to remain intact over an open or opening crack. For the crack-bridging ability of paint in its liquid state, the crack width it is able to bridge should be specified. When set, a crack-bridging paint film will remain intact in case of changes of the crack width. Information of the crack width and crack width changes should be given. Some paint films have the ability to remain intact where a new crack is formed in the substrate. In such cases, specification of the maximum crack width to which the crack may open without cracking of the paint film should be given.

Damage
Defect resulting from the performance of the building.

Defect
See Fault and damage.

Degree of cleaning
Degree of cleaning of reinforcing bars based on visual inspection compared to photographs in ISO 8501-1. There are three categories of cleaning: Sa class with blast cleaning, St class with steel brushing and F class with flame cleaning.

Demolition
Removal of polluted, damaged or sound concrete from a substrate.

Depression
Small (concave) recess in, for example, a ground surface, a road surface, a rock surface or a concrete surface.

Determination of delamination
A mortar wearing course or tile cladding on concrete is considered to be delaminated if a light knock on the surface results in a toneless and coarse sound. This implies that the cladding is delaminated.

Dew point
The temperature at which vapour is condensed into water.

Diffusion
Penetration of a substance into another, e.g. concrete, due to differences in material concentrations where the penetrating substance moves in such a way that the concentration differences tend to be equalized.

Dilation
Property of a material that swells by absorption of water, e.g. swelling injection materials.

Dimensional stability
The ability of a repair product or system when applied to a substrate implying that the repair may absorb stresses due to change of volume.

Elastic
A material is considered to be elastic if, when subjected to a load, it resumes its former shape when the load is removed. In case of proportionality between stress and deformation the material is considered to be linear elastic.

Elastomer
As polymer, but with mechanical properties like rubber, e.g. major elongation at failure.

Electrolyte
An aqueous solution of a substance which is fully or partly ionized and therefore electrically conductive.

Electromigration
Form of transport of ions, where the driving force is an electric difference of potential.

Emulsion
Heterogeneous mix of an atomized liquid into another.

Epoxy
Bonding agent in one- and two-component repair material (e.g. glue, paint, membrane and coating). Available in an aqueous and a non-aqueous form.

Evaluation of state
Conclusion of registration of state and analysis of state.

Execution
The ability of a product or a system to create an efficient and durable protection, repair or strengthening without harmful effects on the concrete structure to which they are applied, or other structures, personnel, users and other persons or the environment.

Expansion joints
Joints which separate a building into sections and which are filled with air or an easily compressible material so that the expansion of a section is not counteracted by the adjacent sections.

Fault
Faulty design or execution.

Finishing treatment
Preparation and finishing of an applied repair material, possibly with subsequent surface protection to ensure that the required properties and characteristics are obtained.

Flame cleaning
Cleaning of a concrete surface by open flame in particular to remove pollution (oil) and surface treatment. The outer millimetres of the concrete are spalling due to the heat as thermocracks are formed. Thermocracks are left in the surface.

Glaze
A transparent paint.

Gravitational injection
Filling of a crack in a horizontal upward concrete surface using gravitation by creating a small pool of the (liquid) filling material over the crack that will then be filled.

High-pressure cleaning or washing
Cleaning with water at a maximum pressure of 300 bars or with air at a maximum pressure of 20 bars.

Hot-water cleaning
Flushing with hot water (below 100 °C) at a maximum pressure of 300 bars.

Hydraulic binder
Cement in accordance with EN 197-1 or EN 413-1, lime according to EN 459-1 or a combination of cement according to EN 197-1, with a type-II additive, which is in conformity with a European standard. Note that all cements and additives should comply with the national standards until such time as the European standards for cements and additives come into force.

Hydraulic injection material
Injection product where the setting is based on hydration of a hydraulic binder.

Hydraulic mortar and concrete
Mortar and concrete with a hydraulic binder mixed with graduated aggregates and often mixed with admixtures and additives. Hydraulic hardening occurs when they are mixed with water.

Hydrophobic impregnation
Product in a liquid state which penetrates the concrete and forms a hydrophobic coat within the concrete voids. As a result, the concrete becomes water-repellent. The voids on the concrete surface and capillaries are not filled and no film is formed. Therefore, the appearance of the concrete surface shows no significant change. The active component of a hydrophobic impregnation may be silane or siloxane.

Identification test
Test method applied to composite materials, products or systems to document their identity through fundamental properties and characteristics. Identification tests are used by the supervisor and the contractor to verify that the applied materials comply with the required concrete specifications. In the EN 1504 standards it is required that the measured properties of products and systems are in conformity with those stated by the manufacturer with further specified tolerances.

Impermeability to water
The ability of a surface protection to prevent water from penetrating through the paint film into the substrate.

Impregnating void-filling material
Product in a liquid state. When applied to a concrete surface it penetrates the voids of the concrete surface and forms a solid material. The voids and capillaries of the concrete surface are partly filled so that a discontinuous, thin film (10 to 100 mm) is formed on the concrete surface. The binder of the film may, for example, consist of organic polymers.

Incipient anode
An area of reinforcement in a corroding concrete structure with original cathodic protection from an adjacent local anode (corroding). Due to repair the anode is removed so that the originally protected area of the reinforcement is no longer protected and starts corroding.

Indicator
Substance that changes colour dependent on the situation. For example, within the concrete field, phenolphthalein is known as a carbonating indicator and silver nitrate is known as a chloride indicator.

Inhibitor
An additive to concrete or mortar which, when added in small quantities, prevents corrosion or inhibits the corrosion rate of corroding reinforcement.

Injection
Filling of cracks and cavities in concrete with a liquid material under pressure (or gravity) with subsequent setting so that density and/or strength are achieved.

Injection products
Products for injection can be classified into three categories according to their intended use:

- Force transmitting injection of cracks, defects and interstices with repair materials which adhere to concrete surfaces in cracks, defects and interstices and thus make transmission of force possible. These are major requirements for the failure criterion, ultimate strain and adhesion in the actual temperature range of the material, and often such injection materials may be brittle after setting.
- Injection materials that are plastic after setting so that cracks, defects and interstices are able to accommodate a subsequent (small) movement.
- Injection materials which are able to swell in connection with setting and water absorption so that cracks, defects and interstices are filled particularly efficiently. Included in this class of injection materials are certain gels which will only seal in constant contact with water. Normally these gels do not bond very well to the substrate.

Inspection
Visual examination of building components for visual imperfections, defects and damage. Concrete samples may be drawn from building components for laboratory tests to determine non-visual imperfections, defects and damage.

Isocyanate
See Polyurethane.

Levelling material
Mortar (putty) for filling irregularities, depressions and cavities less than 3–5 mm deep and open pores in the concrete surface.

Lifetime
The time after the manufacture (e.g. concrete) or positioning (e.g. a building component) where all significant properties and characteristics still fulfil the minimum requirements and the requirements for continuous maintenance are fulfilled.

Low-pressure cleaning and washing
Cleaning with water at a maximum pressure of 100 bars or cleaning with air at a maximum pressure of 7 bars.

Maintenance
Improvements made to restore the original state of a building component provided that it was intended in the design or that it is considered to be necessary based on common practice.

Mechanical decay
Removal of substrate by mechanical action, e.g. a pneumatic hammer.

Membrane
Surface treatment, paint, coating, etc., forming a layer that is impermeable to, for example, water, moisture, chloride and carbon dioxide.

Milling
Method for cleaning and roughening, typically horizontal, upward concrete surfaces.

Monomer
Material consisting of molecules with a low molecular weight and ability to react with the same or other molecules to form a polymer.

Monosilane
A substance which forms silicone when reacting with silicate. Monosilane is normally available as a 40% solution in ethanol (spirits).

Neutral
An aqueous solution is considered to be neutral if its pH value is 7 at 22 °C.

Non-selective waterjet cutting
Removal of substrate by high-pressure waterjet cutting so that sound concrete remains.

Oligomersilane
A type of silicone. It is normally available as a 5–10% solution in white spirit or similar.

Open time
The maximum period of time between mixing of glue and the time at which it is possible to assemble two concrete components so that loading to failure will result in failure of the concrete.

Paint
Liquid material where the chief ingredients are bonding agent, solvent (or water) and pigment. The quantity and type of the bonding agent are crucial for the properties of the paint.

Passive film
Reinforcement is considered to be passive due to a passive film forming on the reinforcement when embedded in concrete. When this passive film is disintegrated (e.g. due to chloride), the reinforcement is no longer protected and will corrode if the necessary conditions are present (water, oxygen).

Passivity
Alternative protection systems are able to reduce the corrosion rate of the reinforcement (to a suitably insignificant degree) provided that these systems are used for their intended purpose which includes paint, plastic coating, metallic coating and inhibitors. The protection obtained by hydraulic cement in concrete is a passivation system.

Performance requirements
The mechanical, physical and chemical properties and characteristics of products and systems required to ensure that protection, repair and strengthening remain durable and stable.

Performance test
Test of properties in which requirements are made for a repair material or system that the properties of the repair material or system are in conformity with the specified requirements at any time during application and use. A performance test is typically made when properties are to be determined for e.g. design purposes. Performance tests and identification tests are often used under the collective name 'test'.

Permeation
Penetration of, for example, water into concrete due to different hydraulic pressure, as the water moves towards the least pressure.

pH value
Negative decimal logarithm for hydrogen ion concentration. Designation for the acidity of an aqueous solution ($pH < 7$ is acid, $pH = 7$ is neutral and $pH > 7$ is basic).

Plastic
A material is considered to be plastic when it is able to change shape when loaded and resume its original shape when unloaded.

Polyester
A type of alkyd, see Alkyd.

Polymer
A material with molecules composed of several, identical recurring units, see Monomers.

Polymer formulated mortar and concrete
Cement-free mortar and concrete with polymer binder (putty), e.g. epoxy, polyurethane or acrylics which set by a polymer reaction.

Polymer injection materials
Injection product for which setting is based on a reactive polymer binder.

Polymer-modified hydraulic mortar and concrete
Hydraulic mortar or concrete modified by addition of polymers. These polymers generally include:

- acrylics, methacrylate or acrylic resins as redispersing powders or aqueous dispersions
- vinyl monopolymers, copolymers and terpolymers as redispersing powders or aqueous dispersions
- styrene–butadiene rubber as an aqueous dispersion
- natural rubber emulsion
- epoxys.

Polyurethane
Addition polymer of a component, containing two or more isocyanate groups, and another component containing two or more hydroxyl groups. In some cases only the name isocyanate is used. There are several solvent-containing and solvent-free polyurethanes as well as aqueous polyurethanes. Polyurethane is used, for example, for glue, injection, paint and binder in some repair mortars. Normally, polyurethane has good ultimate elongation.

Pot life
The period of time taken for a mixed glue, paint, mortar, etc., to reach a specified temperature in the mixing container. Pot life is an identification test carried out under standard laboratory conditions.

Preventive rehabilitation
Rehabilitation to be performed before significant damage can be ascertained.

Primer
Name for priming agent, particularly used for flooring and in connection with surface treatment of steel (reinforcement).

Product
Constituent for protection, repair or strengthening of damaged concrete structures.

Products and systems for adhesion
Products and systems applied to ensure durable bonding between concrete and external reinforcement, e.g. flat rolled steel or carbon fibre strips.

Products and systems for anchorage
Products and systems that:

- anchor reinforcement in concrete to obtain an appropriate structural load-carrying capacity
- fill interstices to ensure proper interaction of reinforcement and concrete.

Products and systems for injection
Products and systems applied for injecting in a concrete structure to restore the structural integrity and/or the durability.

Products and systems for non-structural repair
Products and systems applied for repair of concrete surface layer to rehabilitate the geometry or the aesthetics (but not necessarily the strength).

Products and systems for protection of reinforcement
Products and systems applied to unprotected reinforcement to increase its durability.

Products and systems for structural repair
Products and systems to be used for repair of a damaged surface layer of concrete. Used for replacing the damaged concrete and restoring structural integrity and durability.

Products and systems for surface protection
Products and systems applied to improve the durability of concrete structures (reinforced as well as unreinforced).

Property
A descriptive parameter which is independent of the geometry of the material. Thus, the concrete cube strength is not a property of the concrete, but the cube strength characterizes the strength of the concrete under given circumstances. The cube strength depends on the sideline of the cube, other things being equal.

Protection
Measures with the purpose of preventing water and aggressive substances from penetrating into the concrete of a structure, e.g. impregnation, application of membrane, injection of cracks.

Protection from loading and climate
Protection of newly rehabilitated concrete (repaired, injected and painted concrete) from loading, freezing, driving rain and direct sun, for example.

Quality report
Report on the execution of work, inspection and testing.

RCT
Rapid chloride test. Rapid test method for site measurement of the chloride content of concrete.

Reactive polymer binder
A binder generally consisting of two components: a main component of reactive polymer and a hardener or accelerating agent which is able to harden at ambient temperature. Additives may be applied to a polymer binder. Note that ambient vapour may act as a hardener or an accelerating agent for some systems. Normal polymer binders are, for example:

- synthetic resins (epoxy)
- polyester
- acrylics
- one or two-component polyurethanes.

Recess
Cavity or notch in concrete for fixation of reinforcement or the like.

Reconditioning
New surface protection or repair, typically after climatic exposure for a number of years.

Registration of state
Systematic registration and reporting of the state of, for example, concrete, reinforcement, building components and joints in a concrete structure. Registration of state does not determine any causes of damage (as do Analysis of state and Evaluation of state).

Rehabilitation
General designation for protection, repair and strengthening of concrete structures, i.e. updating a damaged concrete structure.

Removal of cause
Rehabilitation involving or implying removal of the causes of damage.

Repair
Replacing damaged concrete with mortar, concrete or sprayed concrete (shotcrete) to regain strength, density and durability.

Repton
Declaration scheme for cement-based repair materials for concrete.

Roughening of concrete
Removal of substrate to a layer of maximum 15 mm in such a way that the substrate remains rough. The degree of roughness should be specified.

Selective waterjet cutting
Removal of substrate to a predefined depth by high-pressure.

Self-compressing mortar and concrete
A repair product or system, which is proportioned in such a way that extreme yield properties are achieved, and which is able to flow through narrow openings (in moulds) and around closely packed reinforcement without water segregation (bleeding) or aggregates.

Shot blasting
Mechanical method for cleaning and roughening of horizontal, upward concrete surfaces. The method is efficient for large areas and is dust-free.

Shrinkage joint
See Expansion joint and Contraction joint.

Silicone
Product consisting of polysiloxanes. Silicone is available, for example, in the form of rubber (silicone joint filler), resin, impregnation agent or oil.

Sprayed concrete and mortar (shotcrete)
Concrete or mortar applied to substrate at pressure.

Stray current
Reinforced concrete exposed to direct current from, for example, electrically operated tracking vehicles is subject to reinforcement corrosion when the current passes from one conductive bar to another via the concrete so that cathode and anode zones arise (corroding). Stray current does not arise in connection with alternating current.

Strengthening
Measures with the purpose of restoring the original load-carrying capacity of a damaged concrete structure, e.g. by replacing corroded reinforcement (principal reinforcement, transverse reinforcement and stirrups), gluing-on external reinforcement (flat rolled steel, carbon fibre strips, carbon fibre plate) or replacing damaged concrete (in the compressive zone).

Substrate
Concrete to be rehabilitated (base for repair).

Surface protection of concrete
Surface protection (e.g. impregnation, paint, membrane, coating) applied to a concrete surface forms a film, which is able to reduce penetration of water, carbon dioxide and chloride into

concrete. The surface protection may be named water repellent, waterproof, impermeable to water diffusion, and retarding to carbonation and chlorides.

Suspension
A heterogeneous system of a solid matter in a liquid.

Synthetic resin
See Epoxy.

System
Two or more products applied jointly or successively for protection, repair or strengthening of damaged concrete structures.

Technique
Application of a product or a system for which special equipment is required, for example, for injection of cracks (injection technique).

Test of performance characteristics
A test method carried out to verify to the user that a composite material, product or system conforms to its specified performance characteristics or to specify basic physical and mechanical properties and characteristics of, for example, repair materials for design and construction. To be used by the designing engineer for design and construction.

Thermal stability
The ability of a repair product or system which, when applied to a prepared substrate, will have the effect that the concrete is able to sustain cyclic temperature variations.

Thin-layer mortar
See Levelling mortar.

Washing mortar
See Levelling material.

Water atomizing
Continuous freshwater atomizing as a finishing treatment of fresh mortar or concrete.

Water repelling
The ability of paint coatings to prevent water being absorbed and diffused through the paint coat into the concrete substrate.

Water-jet cutting
Removal of concrete by water at a maximum pressure of 1000 bars with 100–200 litres/min or at 2500 bars with 11 litres/min.

Wet-on-wet
Application of mortar/paint in two coatings so that the second coating is applied and worked into the first.

Workability time
The time elapsed from the constituents of a hydraulic or polymer system are mixed and until the mix in the batch quantity used has lost its specified properties and characteristics so that the mix is no longer applicable for its purpose. The workability time is dependent on the batch quantity and is normally also dependent on temperature.

4

Minimum requirements prior to rehabilitation

4.1 General

The following section only deals with the minimum requirements for evaluation of the momentary state of a concrete structure and the possibility that the rehabilitation work will fulfil the significant requirements in, for example, the national standards. The supplementary measures necessary for complying with these requirements and methods when applying the principles, methods, repair materials and repair systems which are described EN 1504 standards, other relevant standards and recognized European technical regulations will be specified.

4.2 Health and safety

Risk to health and safety will be evaluated taking account of the possible event of:

- crashing concrete and rubble
- local failure caused by removal of building material from the concrete structure
- instability of the concrete structure due to removal of damaged concrete and corroded reinforcement.

In such cases where a concrete structure may become unstable in connection with rehabilitation, proper measures will be taken in connection with the design so that the required normative safety is achieved before rehabilitation work is commenced. This will also include the risk involved in the rehabilitation work. Such measures may, for example, include setting up local support and other temporary work, or full or part demolition of the structure or some of the building components.

4.3 Registration and evaluation of state

The damaged concrete structure will be subject to registration and evaluation of damage and defects, their causes and the possibility that the concrete structure can perform as intended. This registration and evaluation of state should be performed by skilled persons with thorough knowledge and practical experience of investigation methods, statistical calculation, maintenance, materials science and such damage mechanisms that may contribute to the disintegration and corrosion of concrete structures. A registration and evaluation of the state of a concrete structure should as a minimum include the following:

- The momentary state of the concrete structure including non-visual and potential damage and defects
- The original basis of design
- Environmental description including action from accumulated pollution
- Special conditions in connection with construction including the climatic action (e.g. winter concreting or dry summer during the construction period)
- History, including overload, rebuilding, etc.
- Use of the structure, e.g. load and other actions
- Requirements for future use of the concrete structure
- The cause or causes why the defects of the damaged concrete structure should be identified and recorded. Many defects are due to faulty design, work specification, construction and materials used. Figures 4.1 and 4.2 give an overview over other frequent causes of defects in concrete and reinforcement corrosion.

Generally, registration and evaluation of state should be performed as a whole and be ready at a suitable time before design of the rehabilitation process is commenced. It is significant that the evaluation is based on the entire damage picture, the causes of damage and the costs.

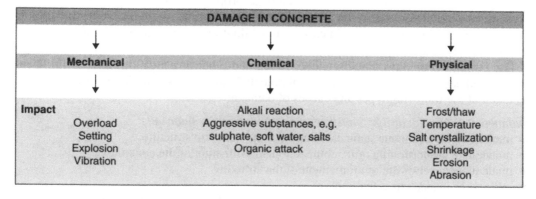

Figure 4.1   General causes of damage to concrete.

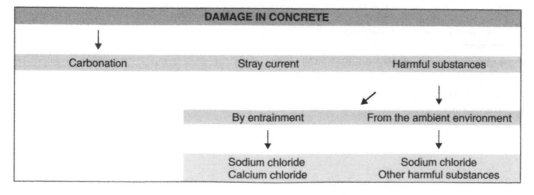

Figure 4.2   General causes of reinforcement corrosion.

A registration and evaluation of the state of a damaged concrete structure should be performed in different phases. The purpose of making a preliminary registration and evaluation of the state may be:

- to get an overview of the state of the concrete structure with regard to its present safety
- to advise the designing engineers and the building proprietor on the significance of performing a detailed registration and evaluation of the state
- to plan the necessary sampling and site investigations of the concrete structure which are necessary to determine the causes and extent of damage.

The purpose of making a detailed registration and evaluation of the state will be:

- to identify the cause or causes of the defects and damage found
- to determine the size and extent of defects and damage
- evaluate whether the defects and damage may be expected to propagate further in the concrete structure to areas where no damage has yet been found (this applies especially to progressing corrosion)
- to estimate the significance of the damage in respect to the normative safety of the concrete structure
- to identify the areas of the concrete structure where rehabilitation (i.e. protection, repair or strengthening) is considered necessary.

An appropriate evaluation of the rate of damage propagation should be given. Then an assessment should be made of the codified remaining lifetime of the concrete structure or the relevant building components in case no rehabilitation other than the established maintenance is performed.

The evaluation of state should include testing and other investigations to reveal hidden defects and damage, as well as any cause of potential damage. As a minimum, account shall be taken of the following types and causes of damage:

Damage due to faulty design, construction and inadequate materials:
- inadequate or erroneous static calculation of the concrete structure
- inadequate proportioning, mix, compaction and finishing of the concrete
- inadequate cover of the reinforcement of the structure
- missing or inadequate dilatation
- insufficient or damaged water protection inclusive of inadequate drainage
- accumulation of harmful substances (e.g. chloride penetration), poor or reactive aggregate (e.g. frost bursting and alkali reaction) due to wrong choice of materials.

Damage due to the serviceability state:
- abrasion
- inadequate and/or faulty maintenance
- collisions
- accumulation of chlorides
- carbonation.

5

Purpose of rehabilitation

5.1 General

This section deals with basic conditions of design, lifetime, costs, maintenance and the extent of the rehabilitation to be considered for the choice of rehabilitation (i.e. protection, repair and strengthening).

5.2 Options

To maintain or restore the codified safety of a damaged concrete structure is a fundamental requirement for any form of rehabilitation. Even when these requirements are met, there will be several options for the rehabilitation.

Consideration in connection with rehabilitation options will normally include investigation of different aspects such as the amount of non-recurrent expenses (e.g. fitting up the building site), the actual costs of rehabilitation and the necessity of restrictions in the use of the structure (e.g. reduced floor load). For the different options, it will probably be expedient to assume different risks of future damage (e.g. demolition and corrosion) to the concrete structure.

The following options may be considered for the choice of adequate measures to fulfil the future requirements for the lifetime of a concrete structure:

- Abstain from immediate rehabilitation and wait for a (specified) period
- Renewed analysis, i.e. an evaluation based on a static calculation, e.g. a probabilistic analysis (probability analysis) of the load-carrying capacity of the concrete structure. This may lead to downgrading of the performance of the concrete structure (e.g. reduced serviceability load)
- Test loading of the building components involved
- Prevent or reduce further decay (e.g. demolition of concrete and corrosion of reinforcement) without actual improvement (i.e. strengthening) of the concrete structure
- Improve or strengthen the entire concrete structure or some of its components
- Replace the entire concrete structure or some of its components
- Remove (demolish) and not replace the entire concrete structure or any of its components.

5.3   Parameters by choice

5.3.1   General

The following considerations for choice of principle and method of rehabilitation are described:

- The intended use of the concrete structure, as well as the required serviceability time and lifetime.
- The performance requirements of the concrete structure. This includes, for example, impermeability and fire resistance.
- The assumed long-term properties and characteristics of the rehabilitation. Maintenance of the rehabilitation work will, however, result in a longer performance period. Lack of maintenance of, for example, joints, drainage and surface protection may thus lead to damage of the concrete structure.
- The possibility, if any, to perform supplementary rehabilitation and instrumentation of the concrete and reinforcement of the structure so that a longer performance period may be obtained for critical damage or so that further propagation of damage can be avoided.
- The number of repeated rehabilitation periods and the incurred costs during the stipulated lifetime. Thus, typically one or more repetitive treatments of a surface coated (e.g. painted) concrete surface. The performance period may be prolonged through proper renewed rehabilitation, but each rehabilitation increases the costs, and certain forms of ageing of the structure will normally continue.
- Costs and considerations of alternative rehabilitation inclusive of the future maintenance and the cost involved.
- Properties and applicability of rehabilitation methods for the existing substrate.
- Appearance and aesthetics of the rehabilitated concrete structure.
- Any inconvenience to the users.
- Access conditions.

5.3.2   Health and safety consequences

In addition to the above conditions, the following consequences to health and safety conditions shall be considered:

- the consequence of failure of the concrete structure or one of the building components (materials and methods)
- health and safety regulations by rehabilitation
- the effect of the rehabilitation work on owners, users and the public.

5.3.3   Structural conditions

In addition to the above conditions, the following structural conditions should be considered:

- possible or necessary changes of dynamic and other structural conditions during and after the rehabilitation
- absorption of loads and other actions during and after the rehabilitation
- necessity of and requirement for future inspection and maintenance.

5.3.4   *Environmental conditions*

In addition to the above conditions, the following environmental conditions should be considered:

- the future environment for the concrete structure and the possibility of local changes
- the need for or the possibility of protecting the concrete structure or some of the building components from the local climate, pollution, chloride-containing traffic splashes, etc., inclusive of protection of the substrate prior to rehabilitation.

5.4   Requirements for choice of principle and method of rehabilitation

Choice of rehabilitation method should comply with the EN 1504 standards and should be based on the following criteria:

- The rehabilitation methods chosen should be efficient for repairing the types of damage, cause or combination of causes that are ascertained.
- The rehabilitation methods chosen should correspond to the future development of the environmental action on the concrete structure.
- The rehabilitation methods should be based on the principles and methods that are specified in section 6 of ENV 1504-9.
- It should be possible to implement rehabilitation methods by using products and systems that are in conformity with EN 1504 standards, other relevant CEN standards or other technical European regulations.

The EN 1504 standards specify requirements for and class of products and systems to be chosen.

6

Choice of repair materials
and systems

6.1 General

This section describes the main principles to be applied individually or in combination in cases where it is necessary to rehabilitate a concrete structure which is subjected to air, is underground or exposed to water (fresh or salt). Thirty-five methods (M1.1 to M11.3) relate to the 11 main principles given in Section 6.2.

Only methods conforming to the main principles should be chosen for rehabilitation of damaged concrete structures. Furthermore, any possible (unintentional) side effect of the actual application should be taken into consideration. These side effects apply to the structure to be repaired and other (adjacent) concrete structures, as well as the personnel and the environment. Then products and systems conforming to the requirements of EN 1504-2 to EN 1504-7 or other European standards should be chosen.

In cases where repair materials and systems are neither covered by an EN 1504 standard nor deviate from the requirements in relevant European standards, the applicability shall be verified in one of the following ways:

- By a European accreditation scheme specially referring to use of the repair materials and systems considered for the rehabilitation according to EN 1504 standards.
- Where there is no European accreditation scheme for the repair materials and systems considered, a relevant national standard or regulation referring to the use of the considered product or system for protection, repair or strengthening according to EN 1504 standards is applicable.

The type of approval certificate issued by Deutsche Institut für Bautechnik, Richtlinien für das Verstärken von Betonbauteilen durch Ankleben von unidirektionalen kohlenstiffaserverstärkten Kunststofflamellen (CKF-lamellen), Typ Sika CarboDur, Zulassungssnummer Z-36.12-29, Berlin, 1997, can be cited as an example.

Another example is the advance approval by the Road Directorate of the surface treatment products of a manufacturer. When applying products with advance approval, however, reception control by identification testing should be carried out in connection with the execution of the work to ensure that the product delivered conforms to the approved product.

Section 7.4 addresses methods which do not require use of products and systems covered by the EN 1504-2 to EN 1504-7 standards.

If consideration of the options given in Section 5 leads to choice of a specific method of rehabilitation, one of the following principles may be applied:

- prevention or reduction of future impairment of the considered concrete structure and aggravation of its damage without improving the concrete (status quo)
- improvement, strengthening or other updating of all building components in the considered concrete structure.

In such cases the choice of method should be performed according to the main principles specified in Sections 6.2.1 and 6.2.2. Products and systems for the purpose concerned should be chosen in consideration of the substrate (i.e. the structural concrete) and evaluation of damage to concrete and reinforcement and their causes should be in accordance with the specifications in ENV 1504-9, 4.3.

After the choice of a principle or a combination of principles for rehabilitation of the concrete structure, products and systems should be chosen accordingly.

The EN 1504 standards specify requirements for these repair materials and systems at three levels:

- It is required that a number of the repair material properties and characteristics are determined and specified in the information material of the manufacturer as the characteristic value or a certified value (see Annex A).
- It is required that these properties and characteristics fulfil specified minimum requirements.
- It is required that certain properties and characteristics of such repair materials conform to certain compliance criteria with given tolerances to be used for reception control (identification control).

In Annex E a description of the different methods (for principles P1 to P11) is given using, for example, the above set of requirements for the different types of repair materials and systems.

6.2   Principles of rehabilitation

The principles described in the following are based on the chemical and physical laws that allow prevention or stabilization of the chemical or physical decay processes in the concrete or electrochemical corrosion of the surface of the steel reinforcement.

The following methods are examples of use of the given principles. Other methods may be applied subject to verification of conformity with the actual principles. Specification of the methods mentioned is given in EN 1504-10. Unintentional consequences of the methods applied under certain, special conditions should always be considered.

It may be necessary (or appropriate) to use several methods in combination for rehabilitation of damaged concrete structures. Thereby the interaction of the methods is increased, which may lead to intentional (but also unintentional) effects, and these shall be evaluated in each individual case. Thus it is possible to evaluate the effect of rehabilitation of the concrete structure on corrosion of the reinforcement, e.g. in connection with encapsulation of moisture in the concrete or increase of the concrete temperature.

As examples of unintentional consequences of the methods (to be carefully evaluated) specified in ENV 1504-9 the following can be mentioned:

- Reduction of the moisture of concrete which will normally increase the carbonation rate of the concrete – other things being equal.

- Surface protection which may encapsulate the water content of the concrete and thus reduce the adhesion to the surface protection or reduce the frost resistance of the concrete.
- Post-tensioning of the concrete which may introduce local tensile stresses in the concrete.
- Electrochemical methods which may cause hydrogen brittleness in certain types of reinforcement, alkali reaction in concrete with potential alkali-reactive aggregate, reduced frost resistance due to encapsulated moisture or, in cases of submerged concrete structures, corrosion in adjacent structures or containers.
- Limitation of oxygen by surface protection or water saturation which will increase the possibility of corrosion if the reinforcement in the protected zone is in electrical (i.e. mechanical) connection with reinforcement in an unprotected zone.

Mutual compatibility of the products and systems for protection, repair and strengthening as well as compatibility with the substrate of the structure is assumed.

Normally, a rehabilitation task is performed in the following way:

- First the types of damage should be identified.
- This is obtained through registration, analysis and evaluation of the state of the structure.
- Then the possible principles of rehabilitation should be determined, see Table 6.1.
- Finally, a method should be chosen from among the possible rehabilitation methods, see Tables 6.2 and 6.3. Technique, economics (and aesthetics) are decisive factors for this choice. For example, lack of electrical/mechanical connection between reinforcing bars will be a significant technical defect when using electrochemical methods. This defect may be remedied, but will put a strain on the economy.
- When the method or methods for rehabilitation have been chosen, the relevant standards should be selected, see Tables 6.2 and 6.3.

Table 6.1   Examples of damage and the principles applicable for rehabilitation

	Principle	
Damage, defects and corrosion	Related to concrete damage	Related to reinforcement damage
Penetration of aggressive substances, e.g. chloride, gases, chemicals	P1, P3, P6	P7, P8, P10
Cracks due to load, shrinkage, temperature, etc.	P1, P4	
Carbonation	P1, P2	P7, P8, P10
Alkali reaction	P2, P3	
Frost/thaw	P2, P3	
Collision, erosion, abrasion, etc.	P3	
Reinforcement corrosion	P3, P4	P7, P8, P9, P10, P11
Inadequate or wrongly placed reinforcement	P4	
Overloaded concrete (in compressive zones)	P4	
Too low surface strength on the concrete surface (pull-off strength)	P5	
For dusted concrete surface	P5	
Lack of concrete cover		P7
Stray current		P10
Incipient anodes		P7, P10
Polluted concrete cover (chloride, carbonation)		P7

- In conclusion, design and choice of products and systems should be performed and a work specification for performing the rehabilitation work is made, including inspection, control, etc.

Figure 6.1 gives an overview of this action plan.

Table 6.2 Overview of principles and methods for rehabilitation of damaged concrete

Principle no.	Definition	Method no.	Description of method (brief)	Standard EN 1504	Annex no.
P1	Protection against aggressive substances	M1.1	Hydrophobic impregnation	-2- and -10	E1
		M1.2	Sealing	-2- and -10	E1
		M1.3	Cover of cracks with local membrane	-9- and -10	E1
		M1.4	Filling of cracks	-5- and -10	E1
		M1.5	Changing a crack into a joint	-9- and -10	E1
		M1.6	Structural shielding and cladding	-9- and -10	E1
		M1.7	Surface protection with paint	-2- and -10	E1
P2	Moisture control of concrete	M2.1	Protection by hydrophobic impregnation	-2- and -10	E1
		M2.2a	Surface protection by sealing	-2- and -10	E1
		M2.2b	Surface protection with paint	-2- and -10	E1
		M2.3	Structural shielding and cladding	-9- and -10	E4
		M2.4	Electrochemical dehumidification	-9- and -10	E5
P3	Replacement of damaged concrete	M3.1	Hand filling with mortar	-3- and -10	E6
		M3.2	Recasting with repair concrete	-3- and -10	E6
		M3.3	Sprayed mortar or sprayed concrete	-9- and -10	E6
		M3.4	Replacement of building components	-9- and -10	E7
P4	Strengthening of building components	M4.1	Replacing/supplementing reinforcement	-9- and -10	E8
		M4.2	Reinforcement in bored holes	-6- and -10	E8
		M4.3	Adhering flatrolled steel or fibre-composite materials as external reinforcement	-4, -9- and -10	E9
		M4.4	Application of mortar or concrete	-4- and -10	E6
		M4.5	Injection of cracks, voids and interstices	-5- and -10	E2
		M4.6	Filling of cracks, voids and interstices	-5- and -10	E2
		M4.7	Post-tensioning with external cables	-4- and -10	E8
P5	Improvement of physical resistance of concrete	M5.1a	Application of wearing surface	-3- and -10	E1
		M5.1b	Application of membrane	-3- and -10	E1
		M5.2	Impregnation of the concrete surface	-2- and -10	E1
P6	Improvement of chemical resistance	M6.1a	Increase of chemical resistance by wearing surface	-3- and -10	E1
		M6.1b	Increase of chemical resistance by membrane	-3- and -10	E1
		M6.2	Increase of chemical resistance by sealing	-2- and -10	E1

Table 6.3 Overview of principles and methods for rehabilitation concrete damaged due to reinforcement corrosion

Principle no.	Definition	Method no.	Description of method (brief)	Standard EN 1504	Annex no.
P7	Restoring reinforcement passivity	M7.1	Increase of cover with mortar/concrete	-3- and -10	E10
		M7.2	Replacement of polluted concrete	-3- and -10	E7
		M7.3	Electrochemical re-alkalization	-9- and -10	E12
		M7.4	Passive re-alkalization, natural diffusion	-5- and -10	E11
		M7.5	Electrochemical chloride extraction	-9- and -10	E13
P8	Increase of the electric resistivity of concrete	M8.1	Limitation of moisture in concrete by surface protection (paint, impregnation) or structural protection (panels)	-2- and -10	E1
P9	Control of cathodic areas of reinforcement	M9.1a	Limitation of oxygen admission by saturation	-2- and -10	E1
		M9.1b	Limitation of oxygen admission by membrane		
P10	Cathodic protection of reinforcement	M10.1a	Passive cathodic protection	-7- and -10	E14
		M10.1b	Active cathodic protection	-7- and -10	E14
P11	Control of anodic areas of the reinforcement	M11.1	Reinforcement protection with sacrificial paint	-9- and -10	E15
		M11.2	Reinforcement protection with barrier paint	-9- and -10	E15
		M11.3	Use of corrosion inhibitors for repair	-9- and -10	E16

Table 6.1 gives an overview of the most common causes of concrete damage and reinforcement corrosion along with the related principles applicable to rehabilitation of the building components concerned. These rehabilitation principles are specified in the following text.

Table 6.2 gives an overview of the methods that may be applicable to rehabilitation of concrete damage on building components. These rehabilitation methods are specified in the following text.

Table 6.3 gives an overview of the methods that may be applicable to rehabilitation of corroded reinforcement of building components. These rehabilitation methods are specified in the following text.

The list of rehabilitation methods is not complete, but contains the most general methods for rehabilitation.

It shall be noted that the given principles and methods may be applied individually or in combination, if precautions are taken to ensure that a combination of the different methods will not introduce new damage.

In the rehabilitation of concrete damage and reinforcement corrosion, precautions should be taken to remove the cause of damage before, during or after the rehabilitation (e.g. stray current).

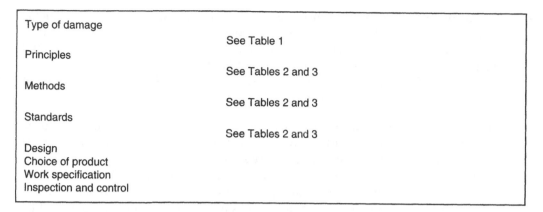

Figure 6.1 Action plan for rehabilitation of concrete damage and reinforcement corrosion.

6.2.1 Principles and methods to prevent defects in concrete

The following principles P1 to P6 cover defects and damage in concrete or concrete structures caused by one or a combination of more of the following situations:

- loading, e.g. by impact, overload, ground subsidence and shrinkage
- chemical and organic action from the environment
- physical action, e.g. frost/thaw, curing heat (thermocracks), moisture migration, pressure from salt crystals and erosion.

In case of further possibility of corrosion, one or more of the principles P7 to P11 should be taken into consideration.

P1: Protection against aggressive substances

Damage and defects may be caused by penetrating substances which are aggressive to concrete and its reinforcement. Typical examples of aggressive substances are chloride (e.g. from seawater, thawing salts, swimming bath water), gases (e.g. carbon dioxide, sulphur dioxide, nitrogen oxide) and chemicals (e.g. acids, bases, fertilizers).

Penetration of the aggressive substances will occur quickly through cracks in the concrete and through the concrete itself if the w/c content is high. Rehabilitation according to the principle in P1 is therefore made by closing these access channels. Prior to this, however, accumulation of harmful substances (to concrete and reinforcement) in the concrete should be removed to an adequate extent.

Cracks. Cracks in reinforced concrete due to structural load are frequent. They may normally be acceptable if their crack width lies within the limit described in ENV 1992-1-1. There may be other causes of cracks, e.g. shrinkage, temperature movements, heat of hydration (thermocracks) and overload. Causes of crack formation and the effect of penetration of aggressive substances have significant influence on the structure and should be subject to evaluation.

The significance of cracks should always be evaluated regardless of their depth and crack width. Cracks may be formed in hardened concrete due to already corroding reinforcement bars. These cracks are normally formed over reinforcing bars and are the first visible sign on onset of corrosion. Cracks over corroding reinforcement cannot be stopped in a simple way by just filling the cracks. Such cracks (due to corrosion) can only be repaired by applying the methods that take account of corrosion, i.e. by using one or more of the principles P7 to P11.

The possibility of future cracking after application of the methods M1.2 and M1.4 should be evaluated, e.g. based on the information given in EN 1504-5.

Methods. ENV 1504-9 lists seven methods based on principle P1 where penetration of aggressive substances into concrete is prevented or retarded to an adequate extent:

- M1.1: *Surface protection by hydrophobic impregnation*. A liquid product, which is applied to the concrete, penetrates the surface layer and changes the surface stress in the concrete voids so that the future penetration of aggressive substances is prevented or retarded.
- M1.2: *Surface protection by sealing*. A liquid product, which is applied to the concrete, forms a continuous film and thus future penetration of aggressive substances is prevented. If there are fine existing or potential cracks in the concrete surface, the paint film should bridge the crack, i.e. able to cover the crack width without bursting. Paint film without proper crack bridging ability should only be applied to crack-free concrete.
- M1.3: *Covering of cracks with local paint*. In case of a single dominating crack in a building component, such a crack may be covered with a local membrane so that penetration into the concrete of harmful substances is prevented. Single cracks are often active (propagating) cracks, which should be taken into account when choosing the type of membrane.
- M1.4: *Filling of cracks*. Cracks above a certain crack width should only be filled or injected so that the crack becomes dense. The method is best applicable to one or a few cracks, but methods for injection of building components with high crack intensity exist. However, these methods are not mentioned in EN 1504 standards. In cases where the building component concerned is a wall cast against ground or rock, injection can be made behind the wall. This is used, for example, in tunnel building in Germany, Switzerland, Sweden and Denmark, but is not mentioned in EN 1504 standards.
- M1.5: *Changing a crack into a joint*. For a crack where the crack width is large and cannot be stopped, changing the crack into a structural joint may be considered. Thereby the joint will also be water-bearing and should be drained through a proper drain in the structure.
- M1.6: *Structural shielding and cladding*. A building component may often be used structurally, e.g. by giving it adequate covering (eaves and panels). Application of this method changes (often drastically) the aesthetics and architecture of the structure.
- M1.7: *Surface protection with paint (membrane)*. Membranes are impermeable and normally also impermeable to water diffusion. Use of membranes for surface protection of building components should be taken into consideration. Membranes are applicable to tightening, but precautions should be taken to prevent the substrate from becoming moist, e.g. by capillary absorption, where the ground water table is placed at a higher level than the bottom of the container. In such cases, adhesion to the substrate may be reduced. Certain polyurethanes are particularly sensitive to this action.

P2: Moisture control of concrete

Development of damage and defects in concrete requires that the concrete is moist. This means that damage and defects do not develop in dry concrete. This is used in principle 2 for moisture control of concrete. Not all building components, however, can be subject to moisture control. Outdoor concrete, especially concrete in water containers, swimming baths, dams, wharfs, outdoor concrete pavements and bridges are some examples of concrete with high moisture content. However, this does not mean that concrete will be damaged or defective if moist. It means that if the conditions for developing damage and defects are present, it will often be possible to stop the propagation of these damages by drying out the concrete. This is, for example, the case for alkali reaction. Carbonation of the concrete cannot be stopped by changing the moisture of concrete. However, the consequence can be removed, as the reinforcement does not corrode at suitably low moisture content which, however, among others depends on the chloride content of the concrete.

If a concrete surface is impregnated, sealed or painted, or if a membrane is applied, the concrete surface will be protected from driving rain. This will reduce the moisture content in the concrete surface, but the concrete will not desiccate. It will normally be regulated to the annual average of the ambient relative humidity. If the building component concerned is in direct contact with the ground and ground water, moisture will be admitted to the concrete by capillary absorption, unless a damp proof membrane is placed, for example, in construction joints.

Alkali reaction. Preconditions for formation of alkali reaction in concrete are:

- presence of alkali reactive particles in the concrete aggregates
- suitably high alkali content in the pore water of the concrete
- admission of moisture to the concrete.

The alkali reaction of concrete can be regulated by controlling the moisture content of concrete. The mechanism is assumed to be such that the alkali-silica gel cannot expand and thus cracks will not be formed, when the concrete is not sufficiently moist for the gel to swell. This situation is assumed when the moisture content of the concrete is less than 70–75% RH dependent on the type of concrete.

When alkali reaction in concrete occurs over many years, crack formation and expansion tend to decrease with time. After such a period, these cracks can be repaired in the same way as passive cracks (see methods M1.4, M4.5 and M4.6). Prior to such repair, it should be checked that the cracks are passive by measuring the crack width and displacement versus time over a sufficiently long period. Cleaning with water may, for example, provoke new alkali reactions.

Carbonation. The carbonation depth in concrete versus time depends on:

- the concrete composition – mainly the w/c-ratio and the type of binder
- concrete density, e.g. compaction
- moisture content and temperature of concrete
- admission of carbon dioxide.

The carbonation of concrete can be controlled (retarded) by controlling the moisture content and temperature of the concrete, as well as limiting the admission of carbon dioxide.

Wet concrete only carbonates slowly – other things being equal. Examples can be found in 100-year-old dams in Norway, where the carbonation depth is but a few millimetres. Dry concrete also carbonates slowly. However, the moisture penetrates deeply to obtain carbonation which corresponds to that of wet concrete. Desiccation of concrete beyond the stage where its steel reinforcement no longer corrodes is not necessary. It should be borne in mind that desiccation can only be established by admitting energy. Shielding of concrete from driving rain will normally increase the carbonation.

The most efficient carbonation delay in concrete is achieved by applying carbonation retarding surface protection (i.e. impregnation, sealing, paint or membrane) to the concrete. The impermeability to water vapour diffusion, however, should be included in the evaluation so that moisture damage does not arise.

Freeze/thaw damage. By limiting the moisture content of concrete, frost damage caused by fresh and saline water can be limited. For frost damage caused by saline water (e.g. seawater), protection from future penetration of chlorides should also be established, see principle P1.

Methods. ENV 1504-9 lists four methods based on principle P2 for limiting the moisture content of concrete to a sufficient degree:

- M2.1: *Surface protection by hydrophobic impregnation.* In this case impregnating liquid is applied to the concrete surface. The voids of the concrete surface do not close, i.e. no film is formed. The treatment results in a water-repellent concrete surface. However, the concrete surface will not be impermeable to high water pressure so that it is not an impermeable treatment as in water containers. For example, concrete surfaces cannot be made impermeable to driving rain at a high wind pressure. A concrete surface given hydrophobic impregnation will normally not change appearance. Typical products for hydrophobic impregnation are silane and siloxane.
- M2.2a: *Surface protection by sealing.* The purpose of this treatment is to reduce the porosity and permeability of the concrete surface. Generally a 10 to 100 mm thick paint film is formed on the concrete surface – thinnest for impregnation and thickest by paint. It is a common feature that both impregnation and paint film can be open to water vapour diffusion. Typical products are organic polymers, bitumen and cement-based paint.
- M2.2b: *Surface protection with paint.* Membranes forming a film of 0.1 to 5.0 mm are normally impermeable to water vapour diffusion. Typical products are organic polymers and organic polymers with cement as filler. Therefore, where moisture migration towards the back of a membrane is possible, moisture may accumulate so that the purpose of the protection is no longer fulfilled and damage and defects may therefore develop, e.g. frost damage. Moisture formed in membrane-covered concrete may also have the effect that the adhesion between membrane and substrate is significantly reduced, and the membrane may therefore lose contact with the surface (typically for membrane of polyurethane).
- M2.3: *Structural shielding and cladding.* In cases where protection against driving rain only is considered adequate, structural cover, shield or cladding may be applied. There will be no active desiccation of the concrete. The moisture content of shielded concrete will normally adapt to the ambient annual average. It should be noted that the carbonation depth of concrete will normally be increased when the concrete is given protection from driving rain. It is not possible to change the environment for concrete by shielding or cladding to passive exposure class (requires insulation and heating). In case of active ventilation, heating and desiccation, the environment can be changed to a passive exposure class. Typical products for shielding are steel plates, bitumen plates and wooden panels.
- M2.4: *Electrochemical dehumidification.* Methods for dehumidification of concrete have been developed by introducing a suitable electric difference of potential. Cyclic stresses are applied to the reinforcement and an electrolytic steel grid is placed on top of the concrete. Also, cyclic stresses are applied to the steel grid. The concrete reinforcement is dominated by positive stress which may give rise to reinforcement corrosion. By applying cyclic negative stress to the reinforcement the tendency to reinforcement corrosion can be countered. The risk of rust attack on the concrete reinforcement shall be taken into account. The method has not gained footing in Denmark and it has not been possible to obtain any convincing effect by laboratory tests.

P3: Replacement of damaged concrete

When damage has developed to such an extent that the concrete has lost its integrity in the area concerned, the damaged concrete can be removed and replaced by new concrete.

Typical of such situations are damage due to collision, frost/thaw, alkali reaction, spalling of concrete layers due to reinforcement corrosion (due to carbonation or chloride penetration) and action from aggressive substances (e.g. sulphate, soft water, acid). In case of accumulation of penetrating aggressive substances it is necessary to obtain a durable repair that a sufficient quantity of the polluted concrete is removed before repair is performed.

Types of repair. EN 1504-3 distinguishes 'structural repair' from 'non-structural repair'. If damaged concrete is restored by new mortar or concrete the repair material of a structural repair should have approximately the same mechanical properties as the adjacent substrate (coefficient of thermal expansion and coefficient of elasticity). If the repair is significantly stiffer than the substrate, the repair will attract stresses from the substrate, and if so, this should be taken into account. If the repair contains corrosive reinforcement the electric properties of the repair material should be approximately the same as for the substrate, otherwise the risk of corrosion due to incipient anodes is present (see principle P11).

For non-structural repairs the coefficient of thermal expansion of the repair material should wherever possible, be less than that of the substrate.

In cases where the substrate is weak, the above conditions should, however, be evaluated and a supplementary method (e.g. impregnation, sealing or paint) should be applied.

To this may be added:

• *aesthetic repairs*, i.e. structural or non-structural repair for which requirements for the appearance of the repair are made.

Requirements for 'aesthetic repairs' may, for example, be requirements for colour and texture. It should be noted that the EN 1504 standards do not specify requirements for aesthetic repairs, but such repairs are just as important to many builders as structural and non-structural repairs.

Types of repair materials. There are the following types of repair mortar and concrete:

• cement-bound mortar and concrete
• polymer-modified cement mortar and concrete
• polymer-bound mortar and concrete.

Polymer-bound mortar and concrete are generally impermeable to water vapour diffusion and should therefore only be applied for special cases, where it can be documented that the impermeability of the repair will not be harmful to the building component. In a building component, where the moisture migrates from the inside of the substrate towards the repair, moisture may accumulate which will result in damage. Usually, it is only possible to use polymer-bound repair materials for small-size repairs.

Usually, the total contraction of polymer-modified cement mortar and concrete is less than for cement-bound mortar and concrete. Contraction of the repair material may result in crack formation. If the tensile strength of the repair material is too high compared to the bond strength on the substrate, a ring-shaped crack may form along the boundary of the repair. If the bond strength on the substrate is high and if the tensile strength of the repair material is low, cracks may develop in the surface of the repair.

To ensure the best possible bonding between repair material and substrate, the cutting surface of the substrate should be cleaned for microcracks (sand blasting and high-pressure washing) and a bond-improving product (i.e. a primer) may be applied.

Methods. ENV 1504-9 lists four methods, based on principle P3, for replacing damaged concrete:

• M3.1: *Hand-filling with repair mortar.* Only be applied to small repairs.

- M3.2: *Recasting with repair concrete*. In case of a large recreation of damaged concrete, recasting with concrete may be performed. Further to complying with the codified requirements for casting, the shrinking aspects should be subject to evaluation, especially where there is risk of shrinkage cracks (encasing of columns). In case of requirements for the texture of the repair, recasting with fluid mortar may be performed, e.g. for repairs on walls and columns. The casting mould is then adjusted to the required surface structure.
- M3.3: *Application of sprayed mortar or sprayed concrete*. For repairs on a large area, sprayed mortar or sprayed concrete may be applied, e.g. to large wall areas or columns. However, it is difficult to obtain an even and smooth texture. Two methods exist, i.e. dry spraying or wet spraying, and both methods have advantages and drawbacks.
- M3.4: *Replacement of building components*. In case of extensive repairs, replacement of the building components concerned will be an economic advantage – especially for pre-cast building components. Such replacement requires that the structure is braced during the replacement. For replacement of building components in monolithic structures, redistribution of stresses may arise, which should be taken into account (e.g. by replacing a floor beam in a box girder).

P4: Strengthening of building components

Damage and defects may be so extensive that the structure concerned should be strengthened to achieve the required codified safety. The most frequent causes for strengthening a concrete structure are, among others:

- *Principal reinforcement*. The cross-section of the principal reinforcement may be reduced due to corrosion, or faulty design may have resulted in insufficient reinforcement so that the codified safety is not ensured. Furthermore, there may be faulty execution, e.g. placing of insufficient reinforcement or wrongly placed reinforcement.
- *Shear reinforcement*. The cross-section of the shear reinforcement may be reduced due to corrosion. Since the shear reinforcement is placed on top of the principal reinforcement, corrosion of shear reinforcement is normally more frequent than corrosion of the principal reinforcement. Furthermore, faulty design may have resulted in insufficient reinforcement to ensure the codified safety. Furthermore, there may be faulty execution, e.g. placing of insufficient reinforcement or wrongly placed reinforcement.
- *Concrete in the compressive zone*. Extensive damage and defects of concrete in the compressive zone may require strengthening to re-establish the codified safety. The total depth may also be falsely specified in the design material.
- *Formation of cracks*. Normally, reinforced concrete structures are designed on the assumption of zero tensile strength in the concrete. However, special circumstances may require local tensile strength of the concrete – especially in unreinforced concrete structures (e.g. basement walls). In such cases formation of cracks may reduce the codified safety and the crack should therefore be injected based on strength criteria (otherwise, cracks are injected or slushed to achieve density and durability).

When applying strengthening systems the local stresses arising should be taken into account, in particular significant stresses that may arise at anchored reinforcement.

Methods. ENV 1504-9 lists seven methods based on principle P4:

- M4.1: *Replacement or supplementing of embedded reinforcement*. The traditional form of strengthening of the tensile zone of beams and slabs is to break up and remove damaged reinforcement and then supplement the tensile reinforcement with a relevant number of reinforcement bars. This reinforcement may be welded to the original reinforcement if the original as well as the supplementary reinforcement is weldable. It is also possible to bond the supplementary reinforcement to the original reinforcement. However, this requires that the necessary anchorage length is present. Normally, there are no problems using rust-proof or acid-proof reinforcement for supplementing corrosive reinforcement if only the concrete can provide the necessary passivating effect.

- M4.2: *Mounting of reinforcement in bored holes*. Strengthening of reinforced concrete structures sometimes requires anchors, bolts or reinforcement, which are embedded in bored holes in the substrate. Cement paste or polymers can be used for the embedment. Due account should be taken of fire action (fire insulation).
- M4.3: *Adhering flat-rolled steel or fibre composite material as external reinforcement*. The load-carrying capacity of a building component can be increased by adhering external reinforcement of flat-rolled steel, carbon fibre strips or carbon fibre sheets. Synthetic resin glue, which is resistant even in aggressive exposure class, can be applied. The failure moment of beams and slabs can be increased by adhering flat-rolled steel or carbon fibre strip in the tensile zone. The shear capacity of concrete beams can be increased by adhering prefabricated L-shaped carbon fibre strips so that they form a clamp at lap joints. Application of carbon fibre sheets makes the concrete surface impermeable to water vapour diffusion. This shall be taken into account in the design.
- M4.4: *Application of mortar or concrete*. Strengthening of the tensile zone of beams and slabs can be achieved by injection of concrete or mortar. Between the applied concrete layer and substrate, shear stresses due to mechanical load and shrinkage will arise. This should be taken into account in the design. Shear stresses may be absorbed by reinforcing the construction joint with embedded anchors or other reinforcing bars in bored holes in the substrate (method M4.2). The anchors and the reinforcement are fastened with a suitable adhesive of cement paste or polymer.
- M4.5: *Injection of cracks, voids and interstices*. Force transmitting filling of cracks, voids and interstices can be performed by pump injection. The object is to recreate the originally required monolithic properties and characteristics. Typical injection materials are acrylic resin, cement epoxy and polyurethane. Cracks exceeding 0.2 mm on the concrete surface are normally injectable. If the crack width in the surface is less than 0.2 mm, a high pump pressure is required dependent on the viscosity and crack width. The moisture of the substrate to be injected is crucial for the choice of injection material. Water-bearing and moist cracks can be injected with a cement paste of ultra-fine cement or polyurethane which does not blister by contact with water. Certain types of epoxy are less sensitive to moisture, but in principle only dry, force-transmitting cracks should be injected with epoxy. The adhesion strength of the injection material to the substrate should exceed the tensile strength of the substrate. This is measured by a 'slant shear-test'. Active cracks should be injected taking full account of the time-dependence of the crack width. Normally, it is impossible for a crack to achieve strength, as well as dilatation. If the purpose of the injection is to restore the tensile strength of the concrete it should be considered whether supplementary structural reinforcement is necessary.
- M4.6: *Filling of cracks, voids and interstices*. Filling of cracks, voids and interstices to obtain density can be performed by filling and gravity injection. The purpose is to restore the originally required density. Typical injection materials are acrylic resin, cement, epoxy and polyurethane. The moisture of the substrate to be injected is crucial for the choice of injection material. Water-bearing and moist cracks can be injected with a cement paste of ultra-fine cement and certain types of polyurethane which swell by

contact with water. Active cracks should be injected taking full account of the time-dependence of the crack width so that density as well as dilatation is achieved.

- M4.7: *Post-tensioning with external cables*. Reinforcement of building components can be performed by setting up tension cables and post-tensioning them.

P5: Improvement of the physical resistance of concrete

Concrete surfaces are required to have adequate strength, for example, when exposed to heavy traffic loads from the wheel pressure of fork-lift trucks (impact and abrasion). Further, dust may be released from existing concrete surfaces in service, which is annoying, for example, to printing works and bookbinders. If concrete surfaces do not have the necessary strength, or if they are dusty, their serviceability can be restored by repairing the concrete surface.

Methods. ENV 1504-9 lists two methods based on principle P5 for rehabilitation of concrete surfaces:

- M5.1a: *Application of wearing surface*. A weak concrete surface can be rehabilitated by placing a suitable wearing surface of mortar. The possibility of formation of shrinkage cracks should be taken into consideration. However, joints should only be placed above joints in the building component or above cracks (method M1.5). The mortar may tend to segregate from the substrate, especially on floor slabs where the substrate may become hotter than the wearing surface, e.g. over a district heating station. The abrasive strength of the wearing surface can be increased by mixing pulverized steel into the mortar or spreading pulverized steel or carborundum directly after the casting, whereby the wearing surface is plastered.
- M5.1b: *Application of membrane*. Another option is to apply a coating or a membrane to the concrete surface, typically an acrylic coating, epoxy and polyurethane membranes and asphalt mastic. These coatings and membranes should be impermeable even though it is documented that water will not intentionally be used in the room. Unintentional action by water may be spillage, leakages, washing of floors, etc. Use of magnetite mortars in connection with spillage of water results in formation of free chloride which, when penetrating into the substrate, may cause extensive corrosion on reinforcement bars.
- M5.2: *Sealing of surface*. Impregnation with water glass and fluates may reduce or remove the emission of dust. The fluation mechanism is when the free lime in the concrete surface is bound by the sealing:

$$2Ca(OH)_2 + MgSiF_6 \rightarrow \, '2CaF_2 + MgF_2 + SiO_2 + 2H_2O.$$

P6: Improvement of the chemical resistance of concrete

Concrete surfaces are required to have an adequate degree of chemical resistance when exposed to aggressive chemicals (EN 206-1). If concrete surfaces do not have the required durability, serviceability can be achieved by rehabilitating the concrete surface.

Methods. ENV 1504-9 lists two methods based on principle P6 for rehabilitation of concrete surfaces:

- M6.1a: *Increase of chemincal resistance by wearing surface*. A weak concrete surface can be rehabilitated by placing a suitable wearing surface of mortar and thus add the necessary resistance to chemically aggressive substances. The possibility of formation of shrinkage cracks should be taken into consideration. However, joints should only be placed above joints in the building component or above cracks (method M1.5). The mortar may tend to segregate from the substrate, especially on floor slabs where the substrate may become hotter than the wearing surface, e.g. over a district heating station. The durability of concrete and mortar should, as a minimum, be as specified in EN 206-1 for chemically aggressive exposure class.

- M6.1b: *Increase of chemical resistance with membrane.* Another option is to apply a coating or a membrane to the concrete surface. Typical membranes are epoxy and polyurethane coatings. It should be documented that the resistance of these coatings and membranes, as a minimum, corresponds to chemically aggressive exposure class as specified in EN 206-1.
- M6.2: *Increase of chemical resistance by sealing.* Impregnation with water glass and fluates, liquid epoxy, etc., increases the chemical resistance of concrete surfaces. The mechanism is that the free lime of the concrete surface is bound by the sealing (method M5.2).

6.2.2 Principles and methods against corrosion damage

The following principles (P7 to P11) cover reinforcement damage due to corrosion caused by one or a combination of the following:

- Physical loss of the protective cover of the reinforcement
- Chemical loss of alkalinity of the reinforcement cover due to reaction with the carbon dioxide of the atmosphere (carbonation)
- Pollution of the reinforcement cover with corrosion boosting substances, i.e. normally chloride. Chloride may, for example, have been one of the constituents (and therefore mixed with the concrete), or chloride may have penetrated the concrete from the ambient environment
- Stray current from adjacent electric installations.

As mentioned, reinforcement corrosion is possible because the cover of the reinforcement is missing (e.g. due to impact or erosion), is of poor quality, is polluted (e.g. with chloride), is carbonated through or is thinner than necessary. However, reinforcement corrosion may also be caused by stray current even though the quality of the cover (thickness and concrete composition) conforms to the requirements in the project specification.

Normally, dealing with corrosion of reinforcement due to chloride penetration is more difficult than dealing with corrosion due to carbonation. The threshold value for chloride in non-saturated concrete may vary from 0.2 to 0.5% of the cement mass of the concrete (corresponding to approximately 0.02 to 0.06% of the dry matter of the concrete) dependent on type of cement, concrete proportioning (w/c-ratio), the origin of the chloride, admission of oxygen, as well as the concrete alkalinity and environment.

Corrosion due to carbonation. For reinforcement protected by a cover which is carbonated partly through (determined by the phenolphthalein test, see EN 13925) and where it is not practical or economical to recreate a full cover of sufficient quality for long-term protection, the cover can be supplemented with a treatment which retards further carbonation (principle P1), and thus prolong the lifetime of the building component. Further, increased protection can be achieved by increasing the resistivity of the concrete by limiting the moisture content of the concrete according to principle P8.

In cases where the passivity of the reinforcement is completely lost, cathodic protection of the reinforcement may be applied according to principle P10.

Where the reinforcement is no longer adequately protected due to carbonation of the concrete and the concrete is to be repaired by recreating the reinforcement cover with repair mortar according to method M7.3, this protection may be supplemented with protection from carbonation according to principle P1. Alternatively, it is possible to recreate the alkalinity of the concrete on the site using the methods M7.3 and M7.4.

Corrosion due to chloride or other corrosive pollution. If a building component is to be rehabilitated by removing chloride-containing concrete and restoring the protection of the concrete according to method M7.2, all chloride-containing concrete should be removed before recasting, because it is possible that any chloride remaining in the concrete may cause incipient anodes (corrosion). If the moisture transport has been changed due to repair, chloride may be accumulated by diffusion.

Polluting substances other than chlorides, including certain water-soluble chemicals, may also cause reinforcement corrosion, but they only occur in industry and agriculture. Harmful concentrations, however, are normally rare in the natural environment.

Acid pollution with, for example, sulphur dioxide, nitrogen dioxide and nitrogen trioxide in the atmosphere may attack concrete (cause disintegration) as well as reinforcement (cause corrosion) in areas of high pollution, e.g. in chimneys of reinforced concrete.

If concrete is no longer adequately protected against corrosion by the surrounding concrete, the corrosion rate may be reduced by electrochemical chloride extraction according to method M7.5 by increasing the resistivity of the concrete by desiccation according to method M8.1 or by cathodic protection (method M10.1).

For corrosion of reinforcement due to penetrating chloride, the cause of the chloride source should be removed, if possible. If not, the recreation of the cover should be supplemented with protection of the concrete surface from chloride penetration according to principle P1.

Cracks due to chloride action, etc. Where it is possible for chloride or other corrosive substances to penetrate the concrete through cracks, these cracks should be sealed, e.g. according to method M1.4.

Rehabilitation of building components. Protection and repair of concrete building components should be performed by the following principles. In cases of reinforcement corrosion, or risk of potential development of corrosion, the building component concerned shall be protected and rehabilitated according to one or more of principles P7 to P11. Damage to the concrete of the building component should be repaired and protected, and it may be necessary to strengthen the concrete according to principles P1 to P6.

Incipient anodes by patch repair. If corrosion damage is caused by chloride attack (and often carbonation of the concrete), renewed corrosion may often occur along the boundary of a traditional patch repair, i.e. just outside the corrosion protection of the reinforcement.

When patch repair is used, the conditions are changed. It is now assumed that the former rust location (the anode) is repaired traditionally and almost all measures are taken to passivate the reinforcement in a certain, limited area (of repair). In this way a difference arises between concrete around the passivated length of reinforcement (now cathode) and the surrounding, original concrete (which will act as anode), since the latter may contain a little more chloride, have a pH value which is somewhat smaller than the one of the repaired area or have larger defect intensity than the patch-repaired area.

Thus, at the transition between repair concrete and the original concrete, an anode (incipient anode) may be formed on the reinforcement with a cathode on the reinforcement of the repair mortar. This phenomenon is intensified if a shrinkage crack is formed along the repair. Therefore, repair mortar should be made with compensation for shrinkage.

P7: Restoring reinforcement passivity

In cases where reinforcing bars corrode because of depassivation, for example, due to insufficient cover (which may be too thin or porous or worn down) or are transformed (e.g. chloride-containing or carbonated), the reinforcement should be passivated to stop or prevent corrosion.

However, reinforcement may corrode due to stray current even though the cover is sufficient under normal circumstances.

The following methods for passivation of reinforcement are only efficient until a certain depth. There is a possibility of local concentration of aggressive substances from the environment and the concrete adjacent to the repair by formation of incipient anodes. This should be taken into account.

Methods. ENV 1504-9 lists five methods based on principle P7 for rehabilitation of corroded reinforcement:

- M7.1: *Increase of cover thickness with mortar or concrete.* If the concrete of the cover is not polluted with chloride or other aggressive substances, but only too thin, the cover thickness may be increased by applying a suitably dense (i.e. carbonation retarding) mortar. It is, however, a condition that adhesion between mortar and substrate can be ensured in the stipulated lifetime.
- M7.2: *Replacement of carbonated or polluted concrete in covers.* Sometimes it may be necessary to replace both concrete and the corroded reinforcing bars (this is often the case for heavily chloride-containing covers). Furthermore, it may be necessary to protect the repaired as well as the adjacent concrete to counteract incipient anodes. According to principle P1, further carbonation or penetration of aggressive substances from the ambient environment shall be regulated by an equivalent cover of proper thickness.
- M7.3: *Electrochemical re-alkalization.* An electrochemical re-alkalization of carbonated concrete is performed by imposing negative voltage on the concrete reinforcement and applying a steel or titanium grid to the concrete surface in a coating of papier mâché, fabric or similar, which is moistened by an alkaline electrolyte (sodium carbonate or calcium hydroxide). A positive voltage (anode) is imposed on the titanium or steel grid. A current of approximately 1 A per m^2 of concrete surface is supplied and over a period of 1 week (dependent of the conditions) the concrete will re-alkalize the reinforcement by electrolysis (OH^- produced by the cathodic reaction) and electro-osmosis where the alkaline electrolyte penetrates the concrete surface (i.e. migrates). The pH value of the pore water of the carbonated concrete is increased and thus passivates the reinforcement. The calcium carbonate content (from carbonation) of the concrete, however, is not changed. When the re-alkalization of the concrete is determined to be satisfactory (by phenolphthalein test), the rehabilitation is finished apart from cleaning of the concrete surface, patch repairs and surface treatment (with carbonation-retarding paint). Certain types of steel are sensitive to voltage imposed on the reinforcement. They may develop hydrogen brittleness. This applies particularly to certain types of prestressed reinforcement. The substrate may be made with alkali-reactive aggregate and low-alkali cement so that no alkali reaction is developed. It is then possible that the electrochemical method increases the alkali content and the pH value around the reinforcement to such degree that (theoretically) a harmful alkali reaction may arise, if the content of the alkali-reactive aggregate is of sufficient quantity. However, no such cases have been experienced in practice, but have been shown by laboratory tests.
- M7.4: *Re-alkalization of carbonated concrete by natural diffusion.* This method requires application of a layer of mortar (or concrete) with pure (high-alkali) cement binding in such a way that the substrate can be re-alkalized by diffusion of hydroxyl ions from the mortar layer to the substrate. This method depends on the terms present in order to obtain

a sufficiently strong diffusion and so that the mortar layer can prevent new carbonation from reaching the reinforcement (through mortar and cover) for the stipulated lifetime. Strong diffusion is conditional upon mortar with open structure, while protection from carbonation requires a dense mortar.

- M7.5: *Electrochemical chloride extraction.* Extraction of chlorides from the concrete is (theoretically) fairly simple. Chlorides are negatively charged ions. Concrete is porous and this porosity is more or less filled with moisture. The negatively charged chloride ions are therefore dissolved in the pore water of the concrete, where they may migrate (free chlorides). However, some chlorides are chemically and physically bound to the putty of the concrete. By the electrochemical method the free (negative) ions are extracted from the concrete by using the fact that the negative chloride ions are repelled by negative voltage and attracted by positive voltage. Therefore, negative voltage (cathode) is imposed on the reinforcement and wet fabric, papier mâché or similar with a grid of titanium (rust-proof) or of steel (corroding) is applied to the concrete surface and a positive voltage (anode) is applied. A current of approximately 1 A per m^2 of concrete surface is supplied, and in approximately 8 weeks the chlorides migrate to the textile or the papier mâché on the concrete surface. When the chloride content of the concrete is verified to be adequately low, the rehabilitation is finished apart from cleaning of the concrete surface, patch repairs and surface treatment (with chloride-retarding paint). This method of electrochemical chloride extraction from concrete is, generally speaking, only possible when the chlorides are located in the cover and close to the reinforcement. If there are major quantities of chloride behind the reinforcement, accumulation with subsequent local migration of chloride may occur – dependent on the conditions. Certain types of steel are sensitive to voltage applied to the reinforcement, as they may develop hydrogen brittleness. This applies especially to certain pressurized reinforcement. Thus, EN 12696 on 'Cathodic protection of steel in concrete' states that 'hydrogen brittleness of high-strength steel may be avoided if the potential of the reinforcing steel is not below $-900\,\text{mV}$ (SCE)', i.e. is not numerically large. The substrate may contain alkali-reactive aggregates and low-alkali cement so that no alkali reaction is developed. In such cases it is possible that the electrochemical method will increase the alkali content and the pH value around the reinforcement to such degree that (theoretically) there is a risk of harmful alkali reaction, if the content of the alkali-reactive aggregate is sufficiently large. However, no such cases have been observed in practice, but have been shown by laboratory tests.

P8: Increase of the electric resistivity of concrete

The corrosion rate may be reduced by increasing the resistivity of concrete, i.e. the electron flow through the concrete between the cathode and anodic areas on corroding reinforcement. This can be achieved by adequate, active desiccation of the concrete. When the resistivity of the concrete is increased, the corrosion current and thereby the corrosion rate will be reduced.

The resistivity of concrete is measured in the unit Ωm. After curing for 28 days, the resistivity of ordinary Portland cement concrete with w/c-ratio between 0.3 and 0.7 will be in the range of approximately 90 to 40 Ωm, see Hetek-report no. 54. The maturity and type of binder are significant to the resistivity of concrete. The resistivity of the Great Belt Bridge is approximately 650–1000 Ωm dependent on curing temperature and alkali content.

Surface protection of the concrete surface, e.g. according to principle P2, is only resistant to penetration of driving rain. Such surface treatment is not resistant to capillary absorption of moisture from the ground and ground water. Normally, if the concrete is not in contact with moist environments, desiccation can be obtained by surface treatment so that the concrete will be balanced with the humidity of the atmosphere, i.e. the annual average of the ambient relative humidity. However, it is a precondition that the surface treatment does not lead to condensation of water vapour in cold areas, which may add moisture to the concrete by capillary absorption.

It should be borne in mind that desiccation may increase the carbonation rate and thus create other problems.

Methods. ENV 1504-9 only lists one method based on the principle P8 for rehabilitation of corroded reinforcement:

- M8.1: *Limitation of moisture in concrete by surface protection.* Reference is made to principle P2 and the methods given concerning details.

P9: Control of cathodic areas of reinforcement

It may be possible to create such terms that the potential cathodic areas of the reinforcement cannot drive an anodic reaction, i.e. that corrosion will not occur. This can be obtained by limiting or stopping admission of oxygen to all potential cathodic areas on the reinforcement.

It is a precondition that the reinforcement in the building component concerned is electrically insulated from (i.e. not in metallic contact with) reinforcement in other building components with admission of oxygen.

Methods. ENV 1504-9 lists two methods based on principle P9 for rehabilitation of corroded reinforcement:

- M9.1a: *Limitation of oxygen admission by saturation.* The method should only be applied when it is possible that the building component concerned is to be fully immersed and when its reinforcement has no electric contact with building components above the water.
- M9.1b: *Limitation of oxygen admission by oxygen impermeable membrane.* It should be noted that encapsulation of building components of reinforced concrete by an oxygen impermeable membrane will only lead to cathodic control while the membrane is effective (i.e. impermeable to oxygen). Further, measures should be taken to ensure that oxygen is not admitted from locations other than the concrete surface (e.g. from the soil or due to cracks in the membrane or other damage).

P10: Cathodic protection of reinforcement

Cathodic protection of the reinforcement of a building component is in particular efficient against corrosion due to chloride and in cases where the damage due to corrosion are not yet so extensive that it is necessary to remove the concrete and reinforce the building component. It is possible to apply a negative field to the reinforcement of a building component so that none of the potential anodic areas of the reinforcement will be able to emit electrons by anodic reaction, i.e. corrode.

Cathodic protection can be established by the following methods:

- M10.1a: *Passive cathodic protection.* The reinforcement of the building component is connected (electrically) with a sacrificial anode of, for example, magnesium, aluminium or zinc placed in an electrolyte with contact to the reinforcement.
- M10.1b: *Active cathodic protection.* The reinforcement of the building component is provided with a negative electric field with external power supply and the concrete surface is provided with an anode (e.g. in the form of titanium grid, electrically conducting paint or an embedded anode).

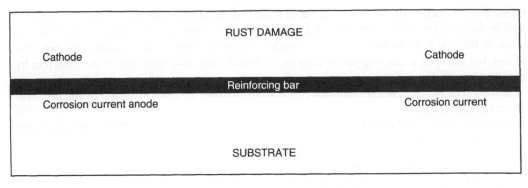

Figure 6.2   The cause of reinforcement corrosion is that the passive cover of the reinforcement is destroyed. This may occur if the carbonation or chloride boundary penetrates deeper than the concrete cover. Thus, when the reinforcement is not protected, random defects and favourable conditions for corrosion (e.g. moisture) may result in corrosion on the reinforcement. Then corrosion will flow from the anode of the reinforcement to the cathode.

Certain types of steel – especially prestressing reinforcement – are sensitive to voltage applied to reinforcements. Therefore, cathodic protection cannot be recommended for prestressing concrete without further investigation, since hydrogen brittleness may develop. This applies especially to certain high-strength steels. Thus, it is stated in EN 12696 on 'Cathodic protection of steel in concrete' that 'hydrogen brittleness of high-strength steel can be avoided if the potential of the reinforcing steel is not lower than $-900\,\text{mV}$ (SCE)', i.e. is not numerically greater.

The substrate may contain alkali reactive aggregates and low-alkali cement so that alkali reaction has not developed. In this case it is possible that the alkali content and the pH value around the reinforcement is increased to such an extent due to cathodic protection that (theoretically) a harmful alkali reaction may occur if the content of the alkali-reactive aggregates is sufficiently large. However, no such cases have been observed in practice, but have been shown by laboratory tests.

The resistivity of repair mortar should correspond to 50–200% of the resistivity of the substrate. Generally, the resistivity of mortar for encasing anodic grids should not exceed $1000\,\Omega\text{m}$. However, it is advisable not to reach this limit, but as a guide, to stick to maximum $100\,\Omega\text{m}$. Besides, EN 12696 on 'Cathodic protection of steel in concrete' should be closely observed.

Acidification at the anode may occur resulting in reduced pull-off strength of the cover.

P11: Control of anodic areas of reinforcement

It is possible to create such conditions for the potential anodic areas of the reinforcement that anode reaction, i.e. reinforcement corrosion, is prevented.

Methods. ENV 1504-9 lists three methods based on principle P11 for rehabilitation of corroded reinforcement:

- M11.1: *Applying reinforcement cover with active pigments (sacrificial paint).* The active pigments may act as active corrosion inhibitors or as sacrificial anodes (e.g. paint rich in zinc). Corrosion inhibitors are chemical additives which counteract formation of anodic

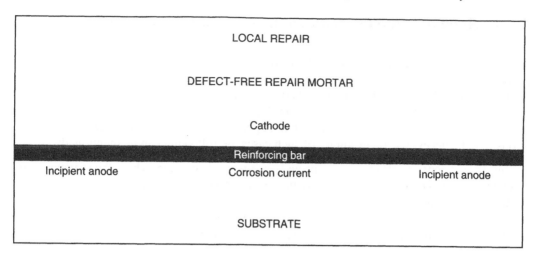

Figure 6.3 By repair the conditions for the existence of the anode are moved (rust area). The former rust area will now act as cathode and incipient anodes may develop at the boundaries of the repair. In case of crack formation, the possibility of incipient anodes is increased.

areas on the reinforcement. There is still some discussion on the long-term effect of corrosion inhibitors. Application of reinforcement coating (paint) containing pigments of lower electric potential than reinforcing steel will have the effect that the cations of the cover are dissolved in the corrosion process instead of the steel of the reinforcement and thus protect the reinforcement against corrosion. The metal ions of the cover may also become active in case of random damage to the reinforcement where the steel is exposed. Other methods include phosphoring with dissolved phosphoric acid and subsequent sealing cover.

- M11.2: *Applying electrically insulating cover (barrier paint) to reinforcement.* If an electrically insulating cover is applied to the reinforcement, emission of ions from the reinforcing steel is prevented so that the corrosion process will not take place. However, the method is only efficient if the reinforcing steel is thoroughly cleaned and the cover is intact (i.e. without any defects). The method shall not be applied, unless the reinforcement can be completely covered along its perimeter. It shall be documented that the anchorage capacity of a surface protected reinforcement is identical to that of unprotected reinforcement. If the anchorage capacity is smaller, this should be taken into account. A typical product is epoxy coating.
- M11.3: *Use of corrosion inhibitors for repair.* Corrosion inhibitors are substances which inhibit reaction or anodic reaction by forming a passive layer on the surface of the reinforcement. Further, during corrosion, inhibitors reduce the corrosion rate. When used for repair, corrosion inhibitors are sprayed on to the cutting surface and the reinforcement prior to application of repair mortar and by mixing them into the repair mortar. Special (organic) corrosion inhibitors applied to concrete surfaces have the ability to migrate in the concrete towards reinforcement to which they are connected. The diffusiveness of concrete to organic corrosion inhibitors is good. Penetration of corrosion inhibitors may also be intensified by electrochemical methods.

6.2.3 Rehabilitation by methods not mentioned in EN 1504

ENV 1504-9 does not contain information of all existing types of repair materials and systems or use of special rehabilitation methods for new and (almost) untried damage. However, it does not necessarily entail that these repair materials and systems are unsuitable or ineffective. Application of rehabilitation methods, which are not empirically well-documented, is therefore not included in EN 1504. This does not mean that the principles on which rehabilitation in ENV 1504-9 are based are not applicable, but that supplementary considerations may be necessary. As examples of such cases the following can be mentioned:

- Reinforced concrete structures – especially prestressed concrete, where the structure and chemistry of the concrete, as well as the tensile strength of cold-drawn reinforcement have been changed due to fire.
- Concrete which has been subjected to high heat exposure through a longish period, for example, concrete in steelworks and concrete in chimneys.

7

Conditional repair materials and systems

7.1 General

In this section it is determined which properties and characteristics repair materials and systems for rehabilitation of concrete structures should possess to conform to the general principles for rehabilitation listed in EN 1504-2 to EN 1504-7, as well as the construction requirements listed in EN 1504-10.

Standards EN 1504-2 to EN 1504-7 and EN 1504-10 specify accepted values for the properties and characteristics of repair materials. When repair methods have been chosen (Section 6), repair materials and systems should be determined in accordance with the requirements of EN 1504-2 to EN 1504-7 and EN 1504-10 and relevant EN standards or technical European regulations. Relevant properties and characteristics are listed in Annex A, Table A.1.

Only repair materials and systems documented by testing according to EN 1504 standards, other relevant EN standards or technical European regulations appropriate for the actual rehabilitation method should be used.

Special account should be taken of the temperature and moisture conditions during repair. Use of most repair materials and systems requires specified intervals for temperature and moisture conditions.

It should be checked that several repair materials and systems applied at the same time will neither be harmful to one another, nor damage the building component concerned.

7.2 Conditions for execution

Repair materials and systems should not be applied under circumstances which are not covered by the relevant EN standards. If the work conditions on a building site are such that the conditions for applying the chosen repair materials and systems cannot be fulfilled (moisture and temperature conditions), other and more suitable materials should be chosen. If this is not possible, other principles and methods fulfilling the rules of application for the rehabilitation concerned should be chosen.

7.3 Required properties of repair materials and systems

Annex A, Table A.1 from ENV 1504-9 specifies the properties and characteristics of repair materials and systems which should be taken into consideration when applying the principles and methods for rehabilitation specified in Tables 6.1–6.3 and Sections 6.2.1 and 6.2.2, respectively.

7.4   Methods not requiring special repair materials or systems

In cases where the methods in Tables 6.2 and 6.3 do not require application of special repair materials or systems, as stated in the EN 1504 standards, relevant requirements for repair materials and systems should be specified.

8

Maintenance after rehabilitation

8.1 General

Unless otherwise specified, for example, in contract or work specification, the following should, as a minimum, be performed when the rehabilitation work is finished:

- Report (description) of the rehabilitation performed, i.e. protection, repair and reinforcement.
- Instructions on the inspection and maintenance to be performed in the stipulated lifetime of the rehabilitated part of the building structure. This applies to, for example, installations for cathodic protection.

In addition, inspection on handing over the work, 1-year inspection and 5-year check-up.

8.2 Report on rehabilitation

In case of a small repair or concrete structure, a detailed report and instruction will normally be adequate (unless otherwise agreed upon) for future maintenance.

In case of a large renovation or more significant structure, the following should be performed, as a minimum:

- Estimation of the remaining lifetime of the concrete structure.
- Identification of the building components whose residual lifetime is expected to be less than the required lifetime for the concrete structure concerned and thus will require renewed treatment at certain intervals. For example, this applied to building components which (according to plan) should be given surface treatment several times in the required lifetime of the structure, because the lifetime of surface protection may be limited. The same applies to joint fillers.
- The dates of the necessary inspection, site testing, laboratory testing and renewed treatment of such building parts.
- Specification of the form of inspection, site testing and laboratory testing to be performed and how to register and treat observations. Finally, a plan for deciding on future inspection of the rehabilitated concrete structure should be specified.
- Specification of continuous maintenance or supervision where this is required, e.g. in connection with cathodic protection (EN 12696).
- Statement of precautionary measures to be taken or prohibitory rules to be introduced concerning the rehabilitation performed, for example, maintenance of drains, maximum pressure at high-pressure washing and application of chloride-containing de-icing salts.

An operating manual, or another system, shall ensure that the planned maintenance after the rehabilitation is in fact performed.

9

Health, safety and environment

A work specification for rehabilitation should comply with the requirements of relevant regulations for health, safety and environment as well as fire safety.

Repair materials and systems for rehabilitation should comply with the requirements of relevant regulations for health, safety and environment, as well as fire safety. This should be fulfilled during manufacture, execution of the work and use.

In case of non-fulfilment for specific repair materials and systems, alternative principles for rehabilitation should be considered so that this conflict is avoided.

Overview of E-annexes with reference to methods

An overview of E-annexes is given below. Each of them is related to one or more methods in EN 1504-9 on 'General principles for the use of products and systems'.

Annex E1:   Surface protection
      M1.1:   Surface protection by hydrophobic impregnation
      M1.2:   Surface protection by sealing
      M1.7:   Surface protection with paint (membrane)
      M2.1:   Surface protection by hydrophobic impregnation
      M2.2a: Surface protection by sealing
      M2.2b: Surface protection with paint
      M5.1a: Application of wearing surface
      M5.1b: Application of membrane
      M5.2:   Sealing of surface
      M6.1a: Increase of chemical resistance by wearing surface
      M6.1b: Increase of chemical resistance with membrane
      M6.2:   Increase of chemical resistance by sealing
      M8.1:   Limitation of moisture in concrete by surface protection
      M9.1a: Limitation of oxygen admission by saturation
      M9.1b: Limitation of oxygen admission by oxygen impermeable membrane

Annex E2:   Injection of cracks, defects and voids
      M1.4:   Filling of cracks
      M4.5:   Injection of cracks, voids and interstices
      M4.6:   Filling of cracks, voids and interstices

Annex E3:   Changing a crack into a dense joint
      M1.3:   Covering of cracks with local paint (membrane)
      M1.5:   Changing a crack into a joint

Annex E4:   Structural protection
      M1.6:   Structural shielding and cladding
      M2.3:   Structural shielding and cladding
      M8.1:   Limitation of moisture in concrete by surface protection

| Annex E5: | Electrochemical dehumidification |
| | M2.4:   Electrochemical dehumidification |

Annex E6:	Repair and replacement of damaged concrete
	M3.1: Hand-filling with repair mortar
	M3.2: Recasting with repair mortar
	M3.3: Application of sprayed mortar or sprayed concrete
	M4.4: Application of mortar or concrete
	M5.1a: Application of wearing surface
	M5.1b: Application of membrane
	M6.1a: Increase of chemical resistance by wearing surface
	M6.1b: Increase of chemical resistance with membrane
	M7.2: Replacement of carbonated or polluted concrete in covers

| Annex E7: | Replacement of building components |
| | M3.4:   Replacement of building components |

Annex E8:	Replacement and supplementing of reinforcement
	M4.1: Replacement or supplementing of embedded reinforcement
	M4.2: Mounting of reinforcement in bored holes
	M4.7: Post-tensioning with external cables

Annex E5:   Electrochemical dehumidification
   M2.4:   Electrochemical dehumidification

Annex E6:   Repair and replacement of damaged concrete
   M3.1:   Hand-filling with repair mortar
   M3.2:   Recasting with repair mortar
   M3.3:   Application of sprayed mortar or sprayed concrete
   M4.4:   Application of mortar or concrete
   M5.1a: Application of wearing surface
   M5.1b: Application of membrane
   M6.1a: Increase of chemical resistance by wearing surface
   M6.1b: Increase of chemical resistance with membrane
   M7.2:   Replacement of carbonated or polluted concrete in covers

Annex E7:   Replacement of building components
   M3.4:   Replacement of building components

Annex E8:   Replacement and supplementing of reinforcement
   M4.1:   Replacement or supplementing of embedded reinforcement
   M4.2:   Mounting of reinforcement in bored holes
   M4.7:   Post-tensioning with external cables

Annex E9:   Reinforcement with fibre composite materials
   M4.3:   Adhering flat rolled steel or fibre-composite material as external reinforcement

Annex E10:   Increase of cover
   M7.1:   Increase of cover thickness with mortar or concrete

Annex E11:   Re-alkalization by natural diffusion
   M7.4:   Re-alkalization of carbonated concrete by natural diffusion

Annex E12:   Electrochemical re-alkalization
   M7.3:   Electrochemical re-alkalization

Annex E13:   Electrochemical chloride extraction
   M7.5:   Electrochemical chloride extraction

Annex E14:   Cathodic protection of reinforcement
   M10.1a: Passive cathodic protection
   M10.1b: Active cathodic protection

Annex E15:   Corrosion protection of reinforcement
   M11.1:   Applying reinforcement cover with active pigments (sacrificial paint)
   M11.2:   Applying electrically insulating cover (barrier paint) to reinforcement

Annex E16:   Corrosion inhibitors
   M11.3:   Use of corrosion inhibitors for repair

Overview of principles and methods with reference to annexes

Description of principle/method		Annex
P1:	Protection against aggressive substances	
	M1.1: Surface protection by hydrophobic impregnation	E1
	M1.2: Surface protection by sealing	E1
	M1.3: Covering of cracks with local paint (membrane)	E3
	M1.4: Filling of cracks	E2
	M1.5: Changing a crack into a joint	E3
	M1.6: Structural shielding and cladding	E4
	M1.7: Surface protection with paint (membrane)	E1
P2:	Moisture control of concrete	
	M2.1: Surface protection by hydrophobic impregnation	E1
	M2.2a: Surface protection by sealing	E1
	M2.2b: Surface protection with paint	E1
	M2.3: Structural shielding and cladding	E4
	M2.4: Electrochemical dehumidification	E5
P3:	Replacement of damaged concrete	
	M3.1: Hand-filling with repair mortar	E6
	M3.2: Recasting with repair concrete	E6
	M3.3: Application of sprayed mortar or sprayed concrete	E6
	M3.4: Replacement of building components	E7
P4:	Strengthening of building components	
	M4.1: Replacement or supplementing of embedded reinforcement	E8
	M4.2: Mounting of reinforcement in bored holes	E8
	M4.3: Adhering flat rolled steel or fibre-composite material as external reinforcement	E9
	M4.4: Application of mortar or concrete	E6
	M4.5: Injection of cracks, voids and interstices	E2
	M4.6: Filling of cracks, voids and interstices	E2
	M4.7: Post-tensioning with external cables	E8
P5:	Improvement of the physical resistance of concrete	
	M5.1a: Application of wearing surface	E1, E6
	M5.1b: Application of membrane	E1, E6
	M5.2: Sealing of surface	E1

Description of principle/method			*Annex*
P6:	Improvement of the chemical resistance of concrete		
	M6.1a:	Increase of chemical resistance by wearing surface	E1, E6
	M6.1b:	Increase of chemical resistance with membrane	E1, E6
	M6.2:	Increase of chemical resistance by sealing	E1
P7:	Restoring reinforcement passivity		
	M7.1:	Increase of cover thickness with mortar or concrete	E10
	M7.2:	Replacement of carbonated or polluted concrete in covers	E6
	M7.3:	Electrochemical re-alkalization	E11
	M7.4:	Re-alkalization of carbonated concrete by natural diffusion	E12
	M7.5:	Electrochemical chloride extraction	E13
P8:	Increase of the electrical resistivity of concrete		
	M8.1:	Limitation of moisture in concrete by surface protection	E1, E4
P9:	Control of cathodic areas of the reinforcement		
	M9.1a:	Limitation of oxygen admission by water saturation	E1
	M9.1b:	Limitation of admission with oxygen impermeable membrane	E1
P10:	Cathodic protection of reinforcement		
	M10.1a:	Passive cathodic protection	E14
	M10.1b:	Active cathodic protection	E14
P11:	Control of the anodic areas of the reinforcement		
	M11.1:	Applying reinforcement cover with active pigments (sacrificial paint)	E15
	M11.2:	Applying electrically insulating cover (barrier paint) to reinforcement	E15
	M11.3:	Application of corrosion inhibitors for repair	E16

Annex A: Characteristic value of observations

According to the EN 1504 standard and the European Concrete Code ENV 1992-1-1, Design of Concrete Structures: General Rules and Rules for Buildings, and EN 206-1, Concrete – Part 1: Specification, Properties, Production and Conformity, the 5-percentile is used as a lower characteristic value and the 95-percentile as an upper characteristic value. To determine the characteristic value based on testing, the following assumptions should be made, see EN 14358:

- The lower characteristic value is defined as the 5-percentile
- The upper characteristic value is defined as the 95-percentile
- Determination of the characteristic value is based on observations from tests at a confidence level of $\alpha = 84.1\%$
- The observations from the test are assumed to be logarithmically normally distributed and statically independent
- The coefficient of variation is unknown.

Calculation of characteristic values based on $n \geq 3$ observations (in practice, however, $n \geq 5$ observations shall be preferred) from the same control section:

$$f_1, f_2, f_3, \ldots, f_n \tag{1}$$

is performed by first determining the average $M_{\ln f}$ of the natural logarithm to (1), i.e. of the values:

$$\ln f_1, \ln f_2, \ln f_3, \ldots, \ln f_n \tag{2}$$

The simplest way of doing this is a spreadsheet, e.g. Excel, see example 1. Next, the lower characteristic value (the 5-percentile) is determined by the following equation:

$$f_{kn} = \exp(M_{\ln f} - k_n \cdot S_{\ln f}) \tag{3}$$

and the upper characteristic value (the 95-percentile) is determined by the following equation:

$$f_{k\phi} = \exp(M_{\ln f} - k_n \cdot S_{\ln f}) \tag{4}$$

Factor k_n is based on the inhomogeneous t-distribution, see EN 14358, and assumes the values detailed in Table A.1 and Figure A.1.

Example 1. The following compression strengths for concrete in a control section have been measured by Capo test: 27.5 25.0 24.5 25.0 22.5 24.0 25.5 28.5 25.0 30.0 MPa.

Calculation of the characteristic value according to eq. (3) can be performed as shown in Table A.2.

In Table A.2 the average and standard deviation of the natural logarithms of the compressive strengths are determined as $M_{\ln f} = 3.24508$, and $S_{\ln f} = 0.08576$, respectively, so that the characteristic value is:

$$f_k = \exp(M_{\ln f} - k_n \cdot S_{\ln f}) = \exp(3.24508 - 2.34 \cdot 0.08576) = 21.00 \, \text{MPa}.$$

Table A.1   Values of the factor k_n in equation (4) (see also Figure A.1)

n	3	4	5	6	7	8	9	10	11	12	15	20	30	50	100
k_n	4.11	3.28	2.91	2.70	2.57	2.47	2.40	2.34	2.29	2.25	2.16	2.07	1.98	1.89	1.81

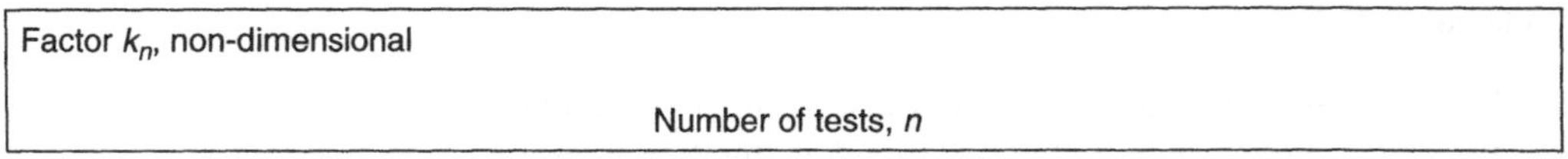

Factor k_n, non-dimensional

Number of tests, n

Figure A.1   Plot of the dependence of the factor k_n and the number of tests n.

Table A.2   Calculation of characteristic value

	Compression strength, f_c MPa	$\ln f_c$
f_c1	27.5	3.14186
f_c2	25.0	3.21888
f_c3	24.5	3.19867
f_c4	25.0	3.21889
f_c5	22.5	3.11352
f_c6	24.0	3.17805
f_c7	25.5	3.23868
f_c8	28.5	3.34990
f_c9	25.0	3.21888
f_c10	30.0	3.40120
Average	25.75	3.24508
Standard deviation	2.252	0.08576
Coefficient of variation	0.087	–
Characteristic value	21.00	–

Annex B: Statistic assessment of identification testing

Introduction

The different standards of the EN 1504 series specify requirements for repair materials and systems in connection with rehabilitation (repair, maintenance and strengthening). If the repair materials and systems used are CE-marked, reception control may be limited to checking the label. If this is not the case, an identification control should be carried out.

Typical requirements for identification control according to an EN 1504 standard are:

- Bilateral requirement for a measurable property of a repair material required in an EN 1504 standard for identification control, e.g. pot life: declared value ±20%.
- Bilateral requirement for a property for which only yes/no description can be given, i.e. colour (the same as the reference?) and location of absorption strip in infrared analysis (the same locations as the reference?).

The purpose of identification control is to check by measuring that a delivery conforms to the expectations based on the manufacturer's information (declaration). Which set of variables to be applied for an identification test is specified in the EN 1504 standards.

As a basis for deciding whether or not identification testing is acceptable, control by alternative measurement, which is independent of the statistic distribution, is applied.

Basis for control by alternative measurement

Control by alternative measurement is based on classification of the samples drawn as 'good' (acceptable) or 'poor' (defective) in relation to the requirements made in the relevant EN 1504 standard. For a delivery to be acceptable, the number of defective samples should not exceed a certain limit in relation to the total number of samples. The minimum number of samples should be three.

If all samples are acceptable, the lot is accepted, and if none of the samples are acceptable, the lot is rejected. If there are both acceptable and non-acceptable samples, acceptance is conditioned upon the size of the lot. For identification testing the number of samples will, however, be small (normally three samples). Therefore, if the lot is to be accepted by the identification test, no non-acceptable samples are allowed.

The basis of the above is given in ISO/DIS 3951-1:2002 on 'Sampling procedures for inspection by variables, Part 1: Specification for single sampling plans indexed by acceptance quality limit (AQL) for lot-by-lot inspection for a single quality characteristic and a single AQL'.

Annex C: CEN and ISO test methods

Table C.1 gives an overview of CEN and ISO test methods in relation to rehabilitation (repair, maintenance and strengthening) of concrete structures.

New test methods related to rehabilitation of concrete are constantly being published and changes are made. Reference is made to the latest edition of the product standard series EN 1504 where the test standards are listed in the section on 'Normative references'. The following test methods are given in numerical order independently of status. Some are available as drafts and some are in preparation. This is not specified in the list, but information of the current status is available at the Danish Standards Association.

Table C.1 Overview of CEN and ISO test standards on rehabilitation

Test no.	Subject
EN ISO 178	Plastics – Determination of flexural properties
EN 196-2	Methods of testing cement – Part 2: Chemical analysis of cement
EN 196-3	Methods of testing cement – Part 3: Determination of setting time and soundness
EN 196-21	Methods of testing cement – Part 21: Determination of the chloride, carbon dioxide and alkali content of cement
EN 445	Grout for prestressing tendons – Test methods
EN ISO 527-1	Plastics – Determination of tensile properties – Part 1: General principles
EN ISO 527-2	Plastics – Determination of tensile properties – Part 2: Test conditions for moulding and extrusion plastics
EN ISO 868	Plastics and ebonite – Determination of indentation hardness by means of a durometer (shore hardness)
EN 1015-3	Methods of test for mortar for masonry – Part 3: Determination of consistence of fresh mortar (by flow table)
EN 1015-4	Methods of test for mortar for masonry – Part 4: Determination of consistence of fresh mortar (by plunger penetration)
EN 1015-6	Methods of test for masonry – Part 6: Determination of bulk density of fresh mortar
EN 1015-7	Methods of test for mortar for masonry – Part 7: Determination of air content of fresh mortar
EN 1015-12	Methods of test for mortar for masonry – Part 12: Determination of adhesive strength of hardened rendering mortars on substrates.

Table C.1   (*continued*)

Test no.	Subject
EN 1015-17	Methods of test for mortar for masonry – Part 17: Determination of water-soluble chloride content of fresh mortars
EN 1062-3	Paints and varnishes – Coating materials and coating systems for exterior masonry – Part 3: Determination and classification of liquid-water transmission rate (permeability)
EN 1062-6	Paints and varnishes – Coating materials and coating systems for exterior masonry – Part 6: Determination of carbon dioxide permeability
EN 1062-7	Paints and varnishes – Coating materials and coating systems for exterior masonry – Part 7: Determination of crack bridging ability
EN 1062-11	Paints and varnishes – Coating materials and coating systems for exterior masonry – Part 11: Methods of conditioning before testing
EN 1081	Resilient floor coverings – Determination of the electrical resistance
EN 1240	Adhesives – Determination of hydroxyl value and/or hydroxyl content
EN 1242	Adhesives – Determination of isocyonate content
EN ISO 1517	Paints and varnishes – Surface-drying test – Ballotini method (also ISO 1517:1973)
EN 1542	Products and systems for the protection and repair of concrete structures – Test methods – Measurement of bond strength by pull-off
EN 1543	Products and systems for the protection and repair of concrete structures – Test methods – Determination of tensile strength – Development for polymers
EN 1544	Products and systems for the protection and repair of concrete structures – Test methods – Determination of creep under tensile strength at $+23\,°C$ and $+50\,°C$ for synthetic resin products
EN 1766	Products and systems for the protection and repair of concrete structures – Test methods – Reference concretes for testing
EN 1767	Products and systems for the protection and repair of concrete structures – Test methods – Infrared analysis
EN 1768	Products and systems for the protection and repair of concrete structures – Test methods – Volatile/non-volatile matter in liquid components
EN 1770	Products and systems for the protection and repair of concrete structures – Test methods – Determination of the coefficient of thermal expansion
EN 1771	Products and systems for the protection and repair of concrete structures – Test methods – Determination of injectability and splitting test
EN 1799	Products and systems for the protection and repair of concrete structures – Test methods – Tests to measure the suitability of structural bonding agents for application to concrete surface
EN 1871	Viscosity
EN 1881	Pull-out test
EN 1877-1	Products and systems for the protection and repair of concrete structures – Test methods – Reactive function related to epoxy resins – Part 1: Determination of epoxy equivalent
EN 1877-2	Products and systems for the protection and repair of concrete structures – Test methods – Reactive functions related to epoxy resins – Part 2: Determination of amine functions using the total basicity number
EN 1878	Products and systems for the protection and repair of concrete structures – Test methods – Thermogravimetric analysis
EN ISO 2409	Paints and varnishes – Cross-cut test (ISO 2509:1992)
EN ISO 2431	Paints and varnishes – Determination of flow time by use of flow cups (ISO 2431:1993, including Technical Corrigendum 1:1994)
ISO 2736-2	Concrete test – Test specimens – Part 2: Making and curing of test specimens for strength tests

Table C.1 *(continued)*

Test no.	Subject
EN ISO 2808	Paints and varnishes – Determination of film thickness (ISO 2808:1997)
EN ISO 2811-1	Paints and varnishes – Determination of density – Part 1: Pyknometer method
EN ISO 2811-2	Paints and varnishes – Determination of density – Part 2: Immersed body (Plummet) method
EN ISO 2812-1	Paints and varnishes – Determination of resistance to liquids – Part 1: General methods
EN ISO 2815	Paints and varnishes – Buchholz indentation test (ISO 2815:1973)
EN ISO 3219	Plastics – Polymers/resins in the liquid state or as emulsions or dispersions – Determination of viscosity using a rotational viscometer with defined shear rate (ISO 3219:1993)
EN ISO 3251	Paints and varnishes – Determination of non-volatile matter of paints, varnishes and binders for paints and varnishes (ISO 3251:1993)
EN ISO 3451-1	Plastics – Determination of ash – Part 1: General methods (ISO 3451-1:1997)
ISO 3951-1	Sampling procedures for inspection by variables – Part 1: Specification for single sampling plans indexed by acceptance quality limit (AQL) for lot-by-lot-inspection for a single quality characteristic and a single AQL
ISO 4628-2	Paints and varnishes – Evaluation of degradation of paint coatings – Designation of intensity, quantity and size of common types of defect – Part 2: Designation of degree of blistering
ISO 4628-4	Paints and varnishes – Evaluation of degradation of paint coatings – Designation of intensity, quantity and size of common types of defect – Part 4: Designation of degree of cracking
ISO 4628-5	Paints and varnishes – Evaluation of degradation of paint coatings – Designation of intensity, quantity and size of common types of defect – Part 5: Designation of degree of flaking
EN ISO 5470-1	Rubber- or plastics-coated fabrics – Determination of abrasion resistance – Part 1: Taber abrader
ISO 6272	Paints and varnishes – Falling-weight test
EN ISO 7783-1	Paints and varnishes – Determination of water-vapour transmission rate – Part 1: Dish method for free films (ISO 7783-1:1996)
EN ISO 7783-2	Paints and varnishes – Coating materials and coating systems for exterior masonry and concrete – Part 2: Determination and classification of water-vapour transmission rate (permeability) (ISO 7783-2:1999)
ISO 8046	Testing of concrete – Hardened concrete – Pull-out strength
ISO 8501-1	Preparation of steel substrates before application of paints and related products – Visual assessment of surface cleanliness – Part 1: Rust grades and preparation grades of uncoated steel substrates and of steel substrates after overall removal of previous coatings
EN ISO 9514 (1992)	Paints and varnishes – Determination of the pot life of liquid systems – Preparation and conditioning of samples and guideline for testing (also ISO 9514:1992)
ENV 10080	Steel for the reinforcement of concrete. Weldable ribbed reinforcement steel B500
ISO 11357-3	Differential scanning calorimetry – Part 3: Determination of fusion temperature
EN ISO 11358	Plastics – Thermogravimetry (TG) of polymers – General principles
EN 12188	Products and systems for the protection and repair of concrete strucures – Test methods – Determination of adhesion steel to steel for characterization of structural bonding agents

Table C.1   (*continued*)

Test no.	Subject
EN 12189	Products and systems for the protection and repair of concrete structures – Test methods – Determination of open time
EN 12190	Products and systems for the protection and repair of concrete structures – Test methods – Determination of compressive strength of repair mortar
EN 12192-1	Products and systems for the protection and repair of concrete structures – Test method – Granulometry size grading – Part 1: Method for dry components of premixed mortar
EN 12192-2	Products and systems for the protection and repair of concrete structures – Test methods – Granulometry size grading – Part 2: Method for fillers for polymer bonding agents
EN 12614	Products and systems for the protection and repair of concrete structures – Test methods – Determination of glass transition temperature of polymers
EN 12615	Products and systems for the protection and repair of concrete structures – Test methods – Determination of slant shear strength
EN 12617-1	Products and systems for the protection and repair of concrete structures – Test method – Shrinkage of polymer binders – Part 1: Determination of linear shrinkage for polymers and surface protecting systems (SPS)
EN 12617-2	Products and systems for the protection and repair of concrete structures – Test methods – Part 2: Volumetric shrinkage
EN 12617-3	Products and systems for the protection and repair of concrete structures – Test methods – Part 3: Determination of early age linear shrinkage for structural bonding agents
EN 12617-4	Products and systems for the protection and repair of concrete structures – Test methods – Part 4: Determination of unrestrained and restrained shrinkage/expansion
EN 12618-1	Products and systems for the protection and repair of concrete structures – Test methods – Part 1: Adhesion and elongation ability of ductile injection products
EN 12618-2	Products and systems for the protection and repair of concrete structures – Test methods – Determination of the adhesion of injection products; with or without thermal cycling – Part 2: Tensile bond method
EN 12618-3	Products and systems for the protection and repair of concrete structures – Test methods – Part 3: Slant shear strength
EN 12636	Products and systems for the protection and repair of concrete structures – Test methods – Determination of adhesion concrete to concrete
EN 12637-1	Products and systems for the protection and repair of concrete structures – Test methods – Part 1: Compatibility with concrete
EN 12637-3	Products and systems for the protection and repair of concrete structures – Test methods – Part 3: Effect of injection on polymer inserts in concrete
EN 12696-1	Cathodic protection of steel in concrete – Part 1: Atmospherically exposed concrete
EN 12715	Execution of geotechnical work – Grouting
EN 12190	Products and systems for the protection and repair of concrete structures – Test methods – Compressive strength
EN 13036-4	Surface characteristics – Test methods – Part 4: Method for measurement of skid resistance of a surface – The pendulum test
EN 13057	Products and systems for the protection and repair of concrete structures – Test methods – Determination of resistance to capillary absorption
EN 13062	Products and systems for the protection and repair of concrete structures – Test method – Determination of thixotropy of products for protection of reinforcement
EN 13294	Products and systems for the protection and repair of concrete structures – Test methods – Determination of stiffening time
EN 13295	Products and systems for the protection and repair of concrete structures – Test methods – Determination of resistance to carbonation

Table C.1   (*continued*)

Test no.	Subject
ISO 13320-1	Particle size analysis – laser diffraction methods – Part 1: General principles
EN 13394	Products and systems for the protection and repair of concrete structures – Test method – Determination of stiffening time
EN 13395-1	Products and systems for the protection and repair of concrete structures – Test methods – Determination of workability – Part 1: Test for flow of thixotropic repair mortars
EN 13395-2	Products and systems for the protection and repair of concrete structures – Test methods – Determination of workability – Part 2: Test for flow of repair grout or mortar
EN 13395-3	Products and systems for the protection and repair of concrete structures – Test methods – Determination of workability – Part 3: Test for flow of repair concrete
EN 13395-4	Products and systems for the protection and repair of concrete structures – Test methods – Determination of workability – Part 4: Application of repair mortar overhead
EN 13396	Products and systems for the protection and repair of concrete structures – Test methods – Measurement of chloride ion ingress
EN 13412	Products and systems for the protection and repair of concrete structures – Test methods – Determination of modulus of elasticity in compression
EN 13501-1	Fire classification of construction products and building elements – Part 1: Classification using test data from reaction to fire test
EN 13529	Products and systems for the protection and repair of concrete structures – Test methods – Resistance to severe chemical attack
EN 13578	Products and systems for the protection and repair of concrete structures – Test method – Compatibility on wet concrete
EN 13579	Products and systems for the protection and repair of concrete structures – Test methods – Drying test for hydrophobic porelining impregnation
EN 13580	Products and systems for the protection and repair of concrete structures – Test methods – Water absorption and resistance to alkali test for hydrophobic porelining impregnation
EN 13581	Products and systems for the protection and repair of concrete structures – Test methods – Determination of loss of mass after freeze-thaw salt stress-testing of impregnated hydrophobic concrete
EN 13584-2	Products and systems for the protection and repair of concrete structures – Test methods – Part 2: Determination of creep in compression
EN 13687-1	Products and systems for the protection and repair of concrete structures – Test methods – Determination of thermal compatibility – Part 1: Freeze–thaw cycling with de-icing salt immersion
EN 13687-2	Products and systems for the protection and repair of concrete structures – Test methods – Determination of thermal compatibility – Part 2: Thunder-shower cycling (thermal shock)
EN 13687-3	Products and systems for the protection and repair of concrete structures – Test methods – Determination of thermal compatibility – Part 3: Thermal cycling without de-icing salt impact
EN 13687-4	Products and systems for the protection and repair of concrete structures – Test methods – Determination of thermal compatibility – Part 4: Dry thermal cycling
EN 13687-5	Products and systems for the protection and repair of concrete structures – Test methods – Determination of thermal compatibility – Part 5: Resistance to temperature shock
EN 13733	Products and systems for the protection and repair of concrete structures – Test methods – Tests to determine the durability of structural bonding agents
EN 13813	Screed material and floor screeds – Properties and requirements of screed materials
EN 13894-1	Products and systems for the protection and repair of concrete structures – Test methods – Determination of fatigue under dynamic loading – Part 1: During cure

Table C.1 (*continued*)

Test no.	Subject
EN 13894-2	Products and systems for the protection and repair of concrete structures – Test methods – Determination of fatigue under dynamic loading – Part 2: After hardening
EN 14038-1	Electrochemical re-alkalization and chloride extraction treatment for reinforced concrete – Part 1: Re-alkalization
EN 14068	Products and systems for the protection and repair of concrete structures – Test methods – Determination of watertightness of injected cracks without movement in concrete
EN 14117	Products and systems for the protection and repair of concrete structures – Test methods – Determination of viscosity of cementitious injection products
EN 14406	Products and systems for the protection and repair of concrete structures – Test methods – Expansion ratio and evolution
EN 14487-1	Sprayed concrete – Part 1: Definitions, specifications and conformity
EN 14497	Products and systems for the protection and repair of concrete structures – Test methods – Determination of the filtration stability
EN 14498	Products and systems for the protection and repair of concrete structures – Test methods – Volume and weight changes by air drying water storage
EN 14630	Products and systems for the protection and repair of concrete structures – Test methods – Determination of carbonation depth in hardened concrete by the phenolphtalein method

Annex E1: Surface protection

Introduction

The standard EN 1504-2 on 'Surface protection systems for concrete' describes repair materials and systems for rehabilitation of damages on building components of concrete by protecting against penetration of aggressive substances, i.e. surface protection, see principle P1 in ENV 1504-9. The protection reduces or prevents penetration of aggressive harmful substances, e.g. water and other liquids, vapour, or the gases, radon, chemicals and biological substances. The purpose of surface protection is to restore a stable condition in the concrete so that it remains durable. The principle P1 (surface protection) is applied in itself or in combination with different methods:

- M1.1:   Surface protection by hydrophobic impregnation
- M1.2:   Surface protection by sealing
- M1.3:   Crack covering with local paint (membrane)
- M1.4:   Filling of cracks
- M1.5:   Changing a crack into a joint
- M1.6:   Structural shielding and cladding
- M1.7:   Surface protection with paint (membrane)
- M2.1:   Surface protection by hydrophobic impregnation
- M2.2a: Surface protection by sealing
- M2.2b: Surface protection with paint
- M2.3:   Structural shielding and cladding
- M5.1a: Application of wearing surface
- M5.1b: Application of membrane
- M5.2:   Sealing of surface
- M6.1a: Increase of chemical resistance by wearing surface
- M6.1b: Increase of chemical resistance with membrane
- M6.2:   Increase of chemical resistance by sealing
- M8.1:   Limitation of moisture in concrete by surface protection
- M9.1a: Limitation of oxygen admission by water saturation
- M9.1b: Limitation of oxygen admission by oxygen impermeable membrane.

Impregnation, sealing or paint is not applied in all the above-mentioned methods to prevent penetration of aggressive substances. Other forms of protection, for example, injection and shielding are used in some methods. The requirements for surface protection differ according to the intended purposes, for example, carbonation retarding surface protection and chloride retarding surface protection. Requirements for products and systems (in the form of impregnation, sealing and paint) and their application are made in EN 1504-2 on 'Surface protection systems for concrete' and EN 1504-10 on 'Site application of products and systems and quality control of the works', respectively.

Reasons for surface protection of concrete

Concrete, which is composed, mixed, cast and given finishing treatment in accordance with recognized proportioning with a target (moderate) lifetime, will normally not require special surface protection. However, there are cases where surface protection of the concrete surface by impregnation, sealing and paint may be necessary to achieve a required (or prolonged) lifetime, for example:

- faulty concrete proportioning
- defects in the concrete work, e.g. insufficient cover of the reinforcement, incomplete compaction or insufficient finishing treatment
- concrete in particularly aggressive environments, e.g. acid action
- concrete exposed to particularly heavy abrasion, e.g. erosion
- concrete required to be impermeable to gases and water, even if unexpected cracks have been detected.

Design of surface protection is a necessary precondition for satisfactory protection of concrete, but a correct and well-documented construction is just as necessary.

Types of surface protection

EN 1504-2 on 'Surface protection systems for concrete' classifies surface protection into three groups:

- Impregnation is performed by a liquid product which penetrates concrete and forms a hydrophobic (i.e. water-repellent) coat internally in the pores of the concrete. This results in increased surface tension and makes the concrete water-repellent. The capillaries of the concrete are not filled and no film is formed (hydrophobic impregnation does not take up any room). There is no significant change to the appearance of the concrete surface. Active components of hydrophobic impregnation may be silane and/or siloxane.
- Sealing is performed by a liquid product with addition of binder which, when applied to a concrete surface, penetrates the pores of the concrete forming a solid material and thus strengthens the surface. The capillaries of the concrete surface are partly filled so that a discontinuous thin film (10–100 mm) is formed on or in the concrete surface (pore blocking, but not all of the pores are blocked). High penetration is called depth impregnation or depth sealing. Sealing may consist of impregnation with more than approximately 20% binder (e.g. organic polymers).
- Paint is defined as liquid material whose main constituents are binder, solvent (including water) and pigment. The type and quantity of the binder are decisive of the properties of the paint. Some paints are permeable to water vapour diffusion. If, on the other hand, the paint is completely impermeable, it is called a membrane.

The concept 'surface protection', however, should include any form of protection of concrete surfaces. Therefore, classification into the following types is convenient:

- Membrane defined as surface treatment, paint, coat or other treatment leaving a layer which is impermeable to, for example, water, moisture, chloride and carbon dioxide. Membranes may be applied in liquid form or as sheets or rolls. The concept of membrane is applied in ENV 1504-9 on 'General principles for application of products and systems' for method M1.7. The concept of membrane is included in standard EN 1504-2 on 'Surface protection systems for concrete' as a special example of paint which is 'absolutely' impermeable (limit value). It should be noted the EN 1504 series does not specify other requirements than those made in EN 1504-2. Often, the builder makes additional requirements, see AAB for concrete bridges.
- Fire protective paint is defined as paint which insulates the substrate in such a way that harmful temperatures due to fire are prevented. The fire protective effect is produced by foaming of the paint thus creating an insulating effect, or by ablation products in the paint (i.e. substances that liberate water). Fire protective paint may, for example, be used for protection of adhered carbon fibre strips.
- Dust sealers defined as sealing or impregnation where the dust particles on concrete surface are chemically or physically bound. Examples of dust binders are treatment with fluates, application of thin epoxy and acrylic plastic sealers.
- Spark-retarding surface treatment defined as an electrically insulating coat which insulate from sparks (static electricity). Typical coats, for example, contain carbon fibres or aluminium powder. To be applied in case of fire hazard.
- Surface protection from organic attack defined as impregnation of concrete surfaces to prevent deposits of moss, algae and lichen, for example, treatment of concrete surfaces with a 6% solution of benzalkonium chloride (note that chloride is here chemically bound and will thus not harm steel reinforcement).
- Rust retarding surface treatment, i.e. impregnation of repaired concrete surfaces with a migrating corrosion inhibitor to reduce the risk of incipient anode effect.

According to Danish traditions, a more detailed division in relation to main constituents and main properties of the surface protection than in EN 1504-2 on 'Surface protection systems' is used in Table E1.1.

Background for methods M1.1, M1.2, M1.3 and M1.7

Concrete in building components subjected to penetration of aggressive substances often shows damage. The reasons for penetration of the aggressive substances may be:

- Cracks in the concrete
- Defects in the concrete (e.g. honeycombs)
- Porous concrete (e.g. too high w/c-ratio).

To restore the damaged concrete to its former state, the concrete is repaired and the concrete surface is given protection against future penetration of aggressive substances. Aggressive substances may be:

- Carbon dioxide resulting in carbonation so that the concrete loses its rust protection (the pH value of the pore water of the concrete is reduced to about 9)
- Water which is necessary for occurrence of corrosion, alkali reaction and frost damage

Table E1.1 Examples of groups of surface protection according to Danish tradition and EN 1504-2

Surface protection	Former Danish designation	EN 1504-2
Oligomer-silane	Impregnation	Hydrophobic impregnation
Silicone resin	Impregnation	Hydrophobic impregnation
Monosilane	Impregnation	Hydrophobic impregnation
Acrylic resin sealing	Sealing	Sealing
Acrylic plastic sealing	Sealing	Sealing
Acrylic plastic paint	Paint	Paint
Acrylic resin paint	Paint	Paint
Acrylic plastic thickfilm paint	Thickfilm paint	Paint
Acrylic modified cement paint	Paint	Paint
Styrene-butadiene modified cement paint	Thin-layer mortar	Paint
Synthetic rubber paint	Paint	Paint
Silicate paint	Paint	Paint
Epoxy thickfilm paint	Paint/membrane	Paint
Polyurethane coat	Coat/membrane	Paint
Acrylic coat	Coat/membrane	Paint

- Chloride which attacks reinforcement and may cause extensive corrosion
- Oxygen which is a precondition for formation of rust.

To achieve a positive result of the rehabilitation, it is important that the polluted concrete is removed before the surface protection is applied. This is particularly important for concrete containing chlorides, while it is not so critical for carbonated concrete, since a natural re-alkalization may occur due to the surface protection (method M7.3).

Cracks

Reinforced concrete normally has cracks. Normally, cracks where the crack width (on the concrete surface) is within the maximum limit (i.e. the maximum allowable crack width) in the Eurocode ENV 1992-1-1 and EN 206-1 or concrete codes valid in the place of use, and cracks, which are formed by a stress field (cracks due to bending, shear and torsion), are acceptable. However, regardless of the crack width, cracks may have damaging effects and should therefore be avoided to the greatest possible extent in the design, and countered by rehabilitation.

However, cracks may be caused by, for example, plastic shrinkage, desiccation shrinkage, thermal contraction, corrosion and overload. Such ('pathological') cracks should be subject to further investigation before rehabilitation, since causes of cracking or the effect of the penetration or diffusion on the structure should be determined prior to rehabilitation and design.

When cracked concrete is rehabilitated by surface protection, changes in the crack width with time are important in the choice of surface protection. Therefore, it is expedient to divide cracks into the following classes:

- Passive ('dead') cracks, i.e. cracks where the width does not change. The cause of crack formation is no longer there, e.g. old shrinkage cracks.

- Active ('live') cracks, i.e. cracks where the crack width is still changing. The cause of crack formation is still there, e.g. durable cracks (alkali reaction, rust formation) and cracks due to varying action (temperature).
- Latent ('dormant') cracks, i.e. cracks which are apparently passive, but may become active by rehabilitation (e.g. due to change of moisture conditions).

Cracks due to ongoing corrosion cannot be stopped by surface protection alone. Therefore, corrosion should be stopped before surface protection is commenced (principles P7 to P11).

Background for methods M2.1, M2.2 and M2.3

Moisture in concrete is a necessary (but not sufficient) condition for the creation of damage, for example, alkali reaction, frost and rust damage. If the concrete does not have the necessary resistance to such damage, it may be appropriate to try to regulate the moisture in the concrete towards a lower level. ENV 1504-9 on 'General principles for application of products and systems' specifies methods to control the moisture content of concrete: M2.1, M2.2 and M2.3.

It should be noted, however, that these methods will result in a slow change of the moisture in the concrete (evaporation from concrete is slow in the Danish climate due to high humidity of the atmosphere), see the description of the method M8.1. If moisture is admitted to concrete structures, for example, by capillary absorption or direct water load (water towers), the concrete will hardly desiccate sufficiently.

Background for methods M5.1, M5.2, M6.1 and M6.2

The physical and chemical resistance of concrete surfaces is not always sufficient for their intended purpose, so that it is necessary to strengthen the concrete surface. Typical examples are concrete surfaces in sewage works, composting tanks, swimming pools and trafficked areas. In case of changes of the use of buildings it may be necessary to strengthen the physical or chemical resistance of concrete surfaces so that the concrete surface will fulfil the requirements.

Methods M5.1 and M6.1

Physical and chemical reinforcement of concrete surfaces can be achieved by application of wearing course or membranes. However, it is a condition that the two layers (wearing course and substrate) can work together. Therefore, a too weak concrete should not only be given a brittle wearing course, but the wearing course should be strengthened in such a way (reinforcement grid or fibre) that any failure sequence will be plastic. In case of very brittle layers on a soft base there is a tendency towards segregation of the tough layers.

If very thick wearing courses are used it may be necessary to anchor them to the substrate to prevent shear failure in the construction joint. Such anchors should be designed in accordance with the concrete codes ENV 1992-1-1 and the standard EN 1504-6.

Skid/slip resistance

When applying the wearing course to trafficked areas, it should be verified that a specified skid/slip resistance of the trafficked area is achieved. The Transport and Road Research Laboratory in the UK has developed measuring equipment, which has been standardized in Germany (Wehner, 1972). The principles of skidproofing are identical, regardless of the type of binder. For a wearing course of concrete as well as applied membranes, skidproofing may be performed in the following ways (with increasing coarseness) subsequent to priming and application of a base course, where the base course of the skidproofing may be flexible for low abrasion intensity. For high abrasion intensity the base course should be inflexible (i.e. tough with toughness shore, $D \geqslant 60$). Skidproofing may be divided into the following groups:

- *Fine skidproofing.* A skidproofing powder is mixed into a membrane mass (for example, of polyurethane, epoxy or acrylics) by (slow) agitation. The powder should be slowly sprinkled over the membrane mass to avoid lumps and air entrainment in the agitation phase. Then the membrane mass is applied, e.g. by rolling. The mix should be stirred very often, since the skidproofing material may settle dependent on the density. It may be difficult to apply the skidproofing material evenly to the floor.
- *Moderate sprinkling of quartz.* A membrane mass (e.g. of polyurethane, epoxy or acrylics) is rolled out. In the wet mass a suitable quantity of quartz grains is lightly sprinkled, for example, by throwing them vertically into the air over the floor. Directly after that, the quartz grains are rolled into the mass. The distribution of skidproofing material may become uneven by, for example, too much rolling.
- *Sprinkling of quartz grains with full saturation.* A membrane mass is applied in a quantity which is suitable for the skidproofing particles chosen. Quartz grains are sprinkled over the wet mass to fully saturate the surface by throwing them vertically up into the air above the floor. Normally this surface should be top sealed.
- *The scraping method.* An additional layer of primer is applied and fully saturated with fine-grained quartz. Then a membrane mass with quartz grains is applied and levelled by a steel scraper which 'rolls' on the largest quartz grains. Note that a rough surface is necessary to apply the scraping method because the steel scraper must 'roll' and not 'slide' on the base coat.
- *Coarse aggregate surface.* A membrane mass (e.g. with fair UV-resistant clear epoxy) is mixed with quartz grains of 4–8 mm. This mass is applied, levelled and compacted, for example, by a steel trowel.
- *Compact floors.* An epoxy coat with a structure much like orange skin made by machine trowelling. Compact floors are easier to clean and keep bacteria-free (in 'negative blisters') than the other skidproofing layers.

The rougher the skidproofing layer the more difficult it is to keep the coating clean. Therefore, skidproofing should not be rougher than necessary. Especially for floors in the food industry and the health sector it is vital for hygiene that floors be kept clean. Reference is made to the rules of the industries concerned, i.e. the rules of the food industry. Bacteria may accumulate in 'negative blisters' and similar defects. Therefore, such floors should be free of 'negative blisters' and similar cavities and defects.

Methods M5.2 and M6.2

Concrete surfaces may be physically as well as chemically strengthened by impregnation with a suitable liquid which is able to penetrate the concrete surface, and harden so that the necessary physical and chemical resistance is created. In the 1970s a method was created by which desiccated concrete (at 120 °C) was saturated by monomers. By subsequent heating of the concrete, polymers were formed (Søpler, 1971). In this way the strength and durability of concrete was significantly improved (Justeness and Søpler, 1992). However, the method was not used in practice, but it illustrates the principle quite well.

Today there are very thin synthetic resins (epoxy) which are suitable for in-depth sealing of (porous) concrete. The penetration depth will depend on the permeability of the concrete (w/c-ratio). Vacuum impregnation is used in some countries to achieve larger penetration depth. Floors emitting dust can be impregnated by a dust binder. The most ordinary impregnation agents are fluates and thin epoxy paint.

The background for method M8.1

Ongoing corrosion (and cracking) can be retarded by desiccation of the concrete, since moisture is a precondition for corrosion. The electrical resistivity of concrete is increased by desiccation. To achieve a favourable effect of desiccation, a very limited ion transport in the remaining pore water of the concrete is assumed.

Limitation of the moisture of concrete is also a precondition for methods M2.1, M2.2 and M2.3, see the section on these methods.

Indoor concrete

The effect of desiccation can, for example, be seen indoors in heated (dry) concrete buildings. Here corrosion is a rare problem even though the reinforcement cover is fully carbonated. The reason is that the low moisture content in closed and heated buildings tends to increase the electric resistivity of the concrete to a level where the corrosion rate is insignificant. There are a number of examples where reinforcement is seen to corrode indoors, if the room is not heated so that the concrete will remain dry.

Outdoor concrete

In certain situations it is possible to create similar conditions outdoors by reducing the moisture content of the concrete (method M8.1), for example, by establishing a ventilated covering, application of water-repellent surface treatment and pore-filling impregnation or surface protection agents. When these protection methods are used it is necessary to take the treatment into careful consideration to prevent moisture due to condensation in cold areas and to ensure that the concrete does not absorb moisture from the ground or become moist due to poor drainage. The possibility of dehumidification should be maintained.

Practical experience in Denmark has shown that surface treatment cannot be assumed to be a means of desiccation of other layers than the surface layer of concrete exposed to heavy, driving rain. A covered, unheated concrete structure will adjust its moisture to the environment. The equilibrium level will depend on the temperature of the structure in relation to its environment. Roughly, in this case the moisture content of the concrete will be equal to the annual average of the relative humidity of the atmosphere. This may explain why in certain (dry) countries good results with desiccation of concrete by surface protection have been achieved, while countries with a humid climate do not have good results.

Theoretically, the principle can be applied, but desiccation of concrete in Denmark is not possible without supply of energy. Thus, it should be assumed that desiccation of a covered or painted concrete structure will only take place if the structure is heated in relation to its surroundings and its former state, for example, transformation of open balconies into 'winter rooms'.

The moisture level at which reinforcement corrosion ceases depends on whether corrosion is due to chloride or carbonation.

Table E1.2   Relative humidity (average) 50 mm below the concrete surface before and after 863 days' protection

Treatment	Type	Moisture before protection % RH	Moisture after protection % RH
Reference	Untreated	93	94
Aluminium panels	Shielding	94	80
Oligomer siloxane/silane	Impregnation	97	81
Isobutyl-tri-ethoxysilane	Impregnation	93	80
Elastic acrylic dispersion	Sealing	90	79

Example

There are only a few investigations from practice into desiccation of outdoor concrete by shielding and surface treatment (impregnation, sealing or paint). One of these investigations (Jensen, 2002), describes the moisture condition in cladding panels of concrete (in the Oslo area) over 863 days. The annual average of the relative humidity in Oslo is approximately 79%.

Aluminium panels were used as shielding and three different surface treatments of the frontage concrete were given. The moisture content of the frontage concrete at a depth of 50 mm below the surface after 863 days is shown in Table E1.2.

It is seen that a desiccation of 11 to 16% RH (point) has been achieved by surface protection compared to the untreated frontage concrete. However, this will not be sufficient to stop corrosion, but will reduce the reaction rate considerably for any ongoing alkali reaction.

Background for the method M9.1

Ongoing corrosion can be notably retarded by establishing a condition so that potential cathodic areas on the reinforcement do not allow anode reaction (formation of rust). This may be performed by limiting the oxygen concentration at the cathode by, for example, water saturation or by an oxygen impermeable membrane.

Method M9.1 assumes that oxygen admission to all potential cathodic areas is prevented to such an extent that anodic areas (corrosion cells) 'suffocate', and corrosion is prevented because the cathodic areas are inactive. It requires that the reinforcement of the building component is fully electrically insulated from adjacent reinforcement which may receive oxygen.

Although the membrane is oxygen impermeable the ongoing corrosion will continue for some time after application of the membrane. This is solely due to consumption of the oxygen which is already inside the concrete (e.g. bubbles and honeycombs).

Tests with encapsulation (carbon fibre plate applied by epoxy glue) of reinforced concrete columns, where the reinforcement corroded, have shown that full encapsulation prevents corrosion considerably. Partial encapsulation, however, had no effect (Debaiky *et al.*, 2001).

'Surface protection' by water saturation

Water saturation of an entire, delimited reinforced concrete building component where the reinforcement is not electrically connected to reinforcement in any other building component to which oxygen is supplied, is an example of method M9.1.

Limitation of oxygen concentration (at the cathode) by water saturation of concrete should only be used in situations where the building is submerged and where the reinforcement in the submerged building component is electrically insulated from any other reinforcement in building parts which are not submerged or where there is no effective return path for ion transport through the concrete.

Surface protection by oxygen impermeable membrane

Limitation of the oxygen admission by application of an oxygen impermeable membrane is one example of limiting the oxygen concentration at all potentially cathodic areas on the reinforcement. Further care should be taken to prevent admission of oxygen from other sources than the concrete surface (e.g. from the soil or through cracks in the membrane, etc.). Method M9.1 only works as long as the membrane is effective (i.e. oxygen impermeable).

A situation where ongoing corrosion can be expected to 'suffocate' may, for example, be corrosion due to carbonation of the concrete layer covering the reinforcement. Full encapsulation can be shown to stop the corrosion. However, it can hardly be excluded that this stop is due to re-alkalization of the concrete around the reinforcement, i.e. corrosion stop by method M7.4: 'Re-alkalization of carbonated concrete by natural diffusion'. The presence of oxygen may retard the measurability of the effect.

Manufacturer/supplier: surface protection

For surface protection certain properties should be fulfilled in order to document the ability to prevent penetration of aggressive substances (Table E1.3), including the possibility of liberating poisonous gases. Furthermore, certain minimum requirements should be fulfilled (Tables E1.4–E1.6). In EN 1504-2 on 'Surface protection systems for concrete' it is required that information of certain properties should always be available for 'all and certain intended uses'. Furthermore, there are certain minimum requirements for these properties.

Table E1.3 divides the required properties of impregnation, sealing and paint into methods in EN 1504-2. This means that requirements for a surface protection may depend on the method to be applied.

Documentation should be presented again in case of crucial changes of the product or terms of manufacture. For example:

- When a new mix design is used or a new type is introduced
- Change of the mix design which may be significant to the properties of the product
- Change of raw materials which may be significant to the properties of the product.

Resistance to penetration of aggressive substances

Surface protection of concrete surfaces is often used to prevent penetration of aggressive substances (CO_2, Cl^-, SO_2, NO_x and H_2O) into building components of reinforced concrete. For design of the necessary surface protection (product and layer thickness) the supplier/manufacturer may use the test methods stated in Table E1.3. However, it seems to be most clear and convenient to specify an equivalent layer thickness (corresponding to a well-defined concrete mix) for the different surface protections.

Table E1.3   Properties and characteristics to be specified for surface protection according to EN 1504-2

Properties and characteristics	Test method	M1.1 with impregnation	M1.2 with sealing	M1.3 with paint	M2.1 with impregnation	M2.2 with paint	M5.1 with paint	M5.2 with impregnation	M6.1 with paint	M8.1 with impregnation	M8.2 with paint
Linear shrinkage	EN 12617-1	~	~	□	~	□	□	~	□	~	□
Compressive strength	EN 12190	~	~	~	~	~	□	~	□	~	~
Coefficient of thermal expansion	EN 1770	~	~	□	~	□	□	~	□	~	□
Abrasion resistance	EN ISO 5470-1	~	~	~	~	~	■	■	~	~	~
Adhesion by cross-cut test	EN ISO 2409	~	~	□	~	□	□	~	□	~	□
Permeability to CO_2	EN 1062-6	~	~	■	~	~	~	~	~	~	~
Permeability to water vapour	EN ISO 7783-1 and 2	~	□	■	~	■	~	~	~	~	■
Capillary absorption and permeability to water	EN 1062-3	~	■	■	~	■	■	■	□	~	■
Diffusion of chloride ions	EN 13396	□	□	□	~	~	~	~	~	~	~
Freeze–thaw cycling with de-icing salt immersion	EN 13687-1	~	□	□	~	□	□	□	□	~	□
Thunder–shower cycling (thermal shock)	EN 13687-2	~	□	□	~	□	□	□	□	~	□
Thermal cycling without de-icing salt immersion	EN 13687-3	~	□	□	~	□	□	□	□	~	□
Ageing 7 days at 70 °C	EN 1062-11	~	□	□	~	□	□	□	□	~	□
Resistance to thermal shock	EN 13687-5	~	~	□	~	~	□	~	□	~	~
Chemical resistance	ISO 2812-1	~	□	□	~	~	~	~	~	~	~
Resistance to severe chemical attack	EN 13529	~	~	~	~	~	~	~	■	~	~
Crack bridging ability	EN 1062-7	~	~	□	~	□	□	~	□	~	□
Impact resistance	ISO 6272	~	~	~	~	~	■	■	~	~	~
Adhesion strength by pull-off test	EN 1542	~	□	■	~	■	■	■	■	~	■
Fire classification of building elements	EN 13501-1	~	~	□	~	□	□	~	□	~	□
Freeze–thaw testing of impregnated surface	EN 13581	□	~	~	□	~	~	~	~	□	~
Slip/skid resistance	EN 13036-4	~	□	□	~	□	□	□	□	~	□
Penetration depth (Table E1.4)	EN 13579	■	■	~	■	~	~	■	~	■	~
Behaviour after artificial weathering, ageing	EN 1062-11	~	~	□	~	□	□	~	□	~	□
Antistatic behaviour	EN 1081	~	~	□	~	□	□	~	□	~	~
Adhesion on wet concrete	EN 13578	~	~	□	~	□	□	~	□	~	~
Water absorption and resistance to alkali tests, impregnation	EN 13580	■	~	~	■	~	~	~	~	■	~
Drying rate for impregnation	EN 13579	■	~	~	■	~	~	~	~	■	~

■, properties and characteristics for all intended uses; □, properties and characteristics for certain intended uses; ~, no requirements.

Table E1.4 Performance requirements for hydrophobic impregnation (of concrete test specimens)

Properties and characteristics	Test method	Requirements
For all intended uses		
Depth of penetration	EN 1766 and EN 13579	Depth of penetration measured on 100 mm concrete test cubes C(0.70) according to EN 1766 (not C(0.45) as given in EN 13579). The depth of penetration is measured with an accuracy of 0.5 mm by breaking of the treated specimen and spraying the fracture surface with water. The depth of penetration should exceed 5 mm
Water absorption and resistance to alkali	EN 13580	Absorption ratio should be less than 7.5% compared with the untreated specimen Absorption ratio after immersion in alkali solution should be less than 10%
Drying rate coefficient	EN 13579	>30%
For certain intended uses		
Chloride diffusion	EN 13396	The coefficient of chloride diffusion only needs to be tested when the capillary absorption of water is greater than 0.01 kg/m^2 h$^{0.5}$. The result to be estimated by the designer
Loss of mass after freeze–thaw action	EN 13581	The loss of mass of the impregnated specimen must occur at least 20 cycles later than that of the non-impregnated specimen. This test is only necessary for structures which may come into contact with de-icing salts

The equivalent concrete thickness of a surface treatment is defined as the thickness of a concrete layer which, at the same difference of concentration, has the same flux as the surface protection. As a characterization of the concrete, the following may be applied:

- w/c-ratio of 0.55 (corresponding to exposure class M) by protection against carbonation
- w/c-ratio of 0.45 (exposure class A) by protection against penetration.

However, the supplier/manufacturer should specify the equivalent concrete thickness of a surface protection in relation to the w/c-ratio, e.g. as a graph. For a film depositing surface protection, the dryfilm thickness should be specified. For a non-film depositing surface protection, the method of application should be specified.

Cover of cracked concrete

Surface protection can be applied to the concrete surface to cover cracked concrete. This may be performed to prevent penetration of aggressive substances, but also for aesthetic reasons, see later. By testing according to the test method EN 1062-7, on 'Crack bridging ability'

Table E1.5   Performance requirements for sealing of concrete surfaces

Properties and characteristics	Test method	Requirements
For all intended uses		
Capillary absorption and permeability to water	EN 1062-3	Less than $0.1 \, kg/m^2 h^{0.5}$
Depth of penetration measured on 100 mm impregnated concrete test cubes C(0.70)	EN 1766 and EN 13579	The depth of penetration is defined with an exactness of 0.5 mm by breaking of the treated specimen and spraying the fracture surface with water. The depth of penetration should be greater than 5 mm
For certain intended uses		
Abrasion resistance (Taber Abrase)	EN ISO 5470-1 or EN 13813 for floors	At least 10% improvement when using abrading wheel with H22 per 1000 cycles at a load of 1000 g in comparison with a non-impregnated concrete surface
Permeability to water vapour	EN ISO 7783-1 and 2	Class I: $s_D < 5 \, m$ (permeable to water vapour) Class II: $5 < s_D < 50 \, m$ (not dense against water vapour and not permeable to water vapour) Class III: $s_D > 50 \, m$ (impermeable to water vapour)
Chloride diffusion	EN 13396 and EN 1062-3	The coefficient of chloride diffusion only needs to be tested when the capillary absorption of water is greater than $0.01 \, kg/m^2 \, h^{0.5}$. The result to be estimated by the designer
Adhesion after thermal compatibility on substrate C(0.70) acc. to EN 1766	EN 1766 and EN 13687-1, 2 and 3	Thermal cycling acc. to EN 13687-1 and 2 is carried out on the same sample After thermal cycling the following should be fulfilled: Visual. No bubbles, cracks and delamination
Outside application with de-icing salt influence (20×) and thermal shock (10×) Outside application without de-icing salt impact (20×)		Pull-off test. Vertical surface Average: $\geqslant 0.8 \, MPa$ and minimum value: $\geqslant 0.5 \, MPa$ Horizontal unloaded surface Average: $\geqslant 1.0 \, MPa$ and minimum value: $\geqslant 0.7 \, MPa$ Horizontal loaded surface Average: $\geqslant 1.5 \, MPa$ and minimum value: $\geqslant 1.0 \, MPa$
Chemical resistance	ISO 2812-1	No visual defects after 30 days' exposure to relevant environments defined in EN 206-1
Impact resistance	ISO 6272	No visual cracks and delamination after loading

Table E1.5   *(continued)*

Properties and characteristics	Test method	Requirements
		Class I: 4 Nm Class II: 10 Nm Class III: 20 Nm
Pull-off testing	EN 1542	Pull-off. Vertical surface Average: 0.8 MPa and minimum value: $\geqslant$0.5 MPa
Concrete C(0.70) acc. to EN 1766, curing 7 days at normal climate and 7 days at 70 °C in comparison with the non-impregnated specimen		Horizontal, unloaded surface Average: 1.0 MPa and minimum value: $\geqslant$0.7 MPa Horizontal, loaded surface Average: 1.5 MPa and minimum value: $\geqslant$1.0 MPa
Resistance to fire	EN 13501-1	Euroclasses
Slip/skid resistance Class I for inside wet surface Class II for inside dry surface Class III for outside application	EN 13036-4	Class I: >40 units wet tested Class II: >40 units dry tested Class III: >55 units wet tested

Table E1.6   Performance requirements for paint on concrete

Properties and characteristics	Test method	Requirements
For all intended uses Abrasion resistance (Taber Abraser)	EN ISO 5470-1 or EN 13813 for floors	Less than 2000 mg lost materials when using abrading wheel with H22 per 1000 cycles and a load of 1000 g
Permeability to CO_2	EN 1062-6	$s_D > 50$ m
Permeability to water vapour	EN ISO 7783-1 and 2	Class I: $s_D < 5$ m (permeable to water vapour) Class II: $5 < s_D < 50$ m (permeable/ not permeable to water vapour) Class III: $s_D > 50$ m (impermeable to water vapour)
Capillary absorption and permeability to water	EN 1062-3	<0.1 kg/m^2 h$^{0.5}$
Resistance to severe chemical attack It is recommended to use testing liquids as specified in EN 13529 Other testing liquids can be agreed between the interested parties	EN 13529	Class I: 3 days without pressure Class II: 28 days without pressure Class III: 28 days with pressure Change of shore D: Less than 50% when measured according to EN ISO 2815 or 24 hours after the paint is removed from chemical attack: EN ISO 868
Impact resistance measured on painted concrete specimens MC(0.40) acc. to EN 1766	ISO 6272	No visual cracks or delamination after testing Class I: 4 Nm

Table E1.6   (*continued*)

Properties and characteristics	Test method	Requirements
The thickness and expected impact load influence the choice of class		Class II: 10 Nm Class III: 20 Nm
Pull-off strength Rigid (brittle) paint with hardness shore D $\geq$ 60 according to EN ISO 868 Concrete MC(0.40) acc. to EN 1766 should be used Curing: 28 days for one-component paint, cement paint and PCC systems, but 7 days for reactive resin systems	EN 1766 and EN 1542	Paint with hardness shore D < 60: Pull-off. Without trafficking Average: $\geq$0.8 MPa and minimum value: $\geq$0.5 MPa Pull-off. With trafficking Average: $\geq$1.5 MPa and minimum value: $\geq$1.0 MPa Paint with hardness shore D $\geq$ 60: Pull-off. Without trafficking Average: $\geq$1.0 MPa and minimum value: $\geq$0.7 MPa Pull-off. With trafficking Average: $\geq$2.0 MPa and minimum value: $\geq$1.5 MPa
For certain intended uses Linear shrinkage Appropriate only for rigid (brittle) paint with application thickness $\geq$3 mm	EN 12617-1	Paint with shore D $\geq$ 60 and thickness $\geq$3 mm: linear shrinkage <0.3%
Compressive strength	EN 12190	Class I: $\geq$35 MPa (for polyamide wheels) Class II: $\geq$50 MPa (for steel wheels)
Coefficient of thermal expansion Only applicable to rigid (brittle) paint with thickness $\geq$1 mm Only for outside application	EN 1770 EN 1766	Paint with hardness shore D $\geq$ 60 and thickness $\geq$1 mm: αT < 30 $\times$ 10^{-6}K^{-1} (for outside application)
Cross cut test measured on coated concrete samples MC(0.40) This test is only for thin smooth films up to 0.5 mm dry thickness	EN ISO 2409 and EN 1766	Cross cut value <GT 2 Distance between cuts: 4 mm Note: The test is carried out in the basic test additionally to the pull-off test. Therefore, on site the cross cut performance test may replace the pull-off test
Chloride diffusion The result should be evaluated by the designer	EN 13396-2	The coefficient of chloride diffusion should only be tested if the capillary absorption of water exceeds 0.01 kg/m^2 h$^{0.5}$
Adhesion after thermal action on the substrate MC(0.40) according to EN 1766 Shore D acc. to EN ISO 868	EN 1766	Thermal cycling acc. to EN 13687-1 and 2 is carried out on the same sample After testing the following should be fulfilled: Visual. No popouts, cracks and delamination
Outside application with de-icing salt influence	EN 13687-1	Paint with hardness shore D < 60: Pull-off. Without trafficking

Table E1.6 (*continued*)

Properties and characteristics	Test method	Requirements
Freeze/thaw cycles with salts (50×) Thunder shower cycling (thermal shock) (10×)	EN 13687-2	Average: ⩾0.8 MPa and minimum value: ⩾0.5 MPa Pull-off. With trafficking
	EN 13687-3	Average: ⩾1.5 MPa and minimum value: ⩾1.0 MPa
Outside application without de-icing salt influence		Paint with hardness shore D ⩾ 60: Pull-off. Without trafficking
Thermal cycling without de-icing salt impact (20×)	EN 1062-11	Average: ⩾1.0 MPa and minimum value: ⩾0.7 MPa
	EN 13687-5	Pull-off. Without trafficking
Inside application Curing: 7 days at 70 °C Resistance to thermal shock (1×)		Average: ⩾2.0 MPa and minimum value: ⩾1.5 MPa
Chemical resistance	ISO 2812-1	No visual defects after 30 days' exposure to relevant environment as defined in EN 206-1
Crack bridging ability[a]	EN 1062-7	The required classes and test conditions are given in Tables E1.7 and E1.8. After testing according to the required class no failures may occur
Resistance to fire	EN 13501-1	Euroclasses
Skid/slip resistance Inside wet surface: Class I Inside, dry surface: Class II Outside dry/wet surface: Class III	EN 13036-4	Class I: >40 units wet tested Class II: >40 units dry tested Class III: >55 units wet tested
Gloss and colourfastness Exposure to UV radiation and humidity only applies to outside surface protection. It should be tested only white and RAL 7030	EN 1062-11	After 2000 hours of artificial weathering No blistering, ISO 4628-2 No blistering, ISO 4628-4 No flaking, ISO 4628-5 Slight colour change, loss of gloss and chalking may be acceptable, but must be described
Antistatic behaviour	EN 1081	Class I: >104 and <106 Ω Class II: >106 and <108 Ω
Adhesion on wet concrete, tests on substrate MC(0.40) This test is only relevant for paint intended to be applied on fresh and humid concretes	EN 13578	After loading: Visual. No blistering, ISO 4628-2 No cracking, ISO 4628-4 No flaking, ISO 4628-5 Pull-off ⩾1.5 MPa and >50% as concrete rupture

[a]Conditioning according to EN 1062-11: Reactive resin systems, 7 days at 70 °C; Dispersion paint at UV radiation and humidity.

The required crack bridging ability should be determined by the designer with respect to local conditions (climate, crack widths and crack movement).

Table E1.7   Test conditions (method A) for static crack bridging rate according to EN 1062-7

Class	Width of bridged crack, mm	Crack opening speed, mm/min
A1	>0.10	~
A2	>0.25	0.05
A3	>0.50	0.05
A4	>1.25	0.5
A5	>2.50	0.5

Test temperature for the classes A2 to A5 is $-10\,°C$. Test temperature for Class A1 is $+21\,°C$. Other test temperatures can be agreed between the interested parties, e.g. $+10\,°C, 0\,°C, -20\,°C, -30\,°C$ or $-40\,°C$. The test temperature should be included in the test result, e.g. A4 ($-20\,°C$).

Table E1.8   Test conditions (method B9) for dynamic crack bridging according to EN 1062-7

Class	w_o, mm	w_u, mm	n	f, Hz	ΔwL, mm	Δw, mm
B1	0.15	0.10 trapezoid	100	0.03	~	0.05
B2	0.15	0.10 trapezoid	1000	0.03	~	0.05
B3.1	0.30	0.10 trapezoid	1000	0.03	~	0.20
B3.2	0.15	~	20 000	1.00	±0.05 sinus	~
B4.1	0.50	0.20 trapezoid	1000	0.03	~	0.30
B4.2	0.30	~	20 000	1.00	±0.05 sinus	~

As test temperature for classes B1 to B4.2, $-10\,°C$ is recommended in EN 1062-7. Other test temperatures can be agreed between the interested parties, e.g. $+10\,°C, 0\,°C, -20\,°C, -30\,°C$ or $-40\,°C$. The test temperature should be included in the test result, e.g. B3.1 ($-20\,°C$). f, frequency; n, number of crack cycles; Δw, change in crack width; ΔwL, load-dependent movement; w_o, maximum crack width; w_u, minimum crack width.

(Tables E1.7 and E1.8), the supplier/manufacturer can document the ability of paint to bridge active cracks. Film depositing paint is divided into five classes (A1 to A5) and the capability of paint is then stated as, for example, A4 ($-20\,°C$). This means that the paint has the capability to bridge a crack width of 1250 μm at a rate of 0.5 mm/min at temperature $-20\,°C$ according to method A in EN 1062-7.

Generally, paint is applied in several layers each with different properties (e.g. primer, paint and topcoat). It is the property of the total surface protection which is important. In surface protection which, for example, consists of elastic washing mortar with covering paint, the washing mortar will be able to bridge a larger crack width than the paint alone.

For major changes of crack width, paint (and membranes) may be reinforced by elastic textile and thus achieve improved capacity to bridge cracks.

Aesthetic surface treatment

Many concrete surfaces are given surface protection for aesthetic reasons only (cracked and damaged surfaces will reduce the market value of the building). For surface treatment performed solely for aesthetic reasons, special requirements for keeping up the appearance are made, for example:

- weather-resistance, colourfastness and maintenance of the original properties and characteristics
- dirt-repellent, i.e. the ability not to be smudged by soil particles, pollen, algae and soot. The workmanship may, however, also influence the dirt-repellent ability, for example, if the paint is performed with texture. The dirt-repellent ability would be different if the latest treatment were performed with vertical or horizontal brush strokes
- abrasion strength, i.e. the ability to resist mechanical action such as wind, rain, games, traffic load, graphite and cleaning
- aesthetics, including gloss, nature of surface and colour.

It should be noted that EN 1504-2 on 'Surface protection systems for concrete' does not include requirements for the above properties and characteristics.

Designer/builder: surface protection

Surface protection of a building component of concrete should comply with certain requirements to be a suitable protection against penetration of aggressive substances. In EN 1504-2 on 'Surface protection systems for concrete' minimum requirements are made for surface treatment in the form of impregnation, sealing and paint, all of which should be documented.

For design purposes, the properties of surface protection should be given as characteristic values (5-percentile for strength, and 95-percentile for the coefficient of chloride diffusion) or as declared values with very little probability of being exceeded.

Protection against penetration of aggressive substances

First the degree of the required protection should be determined. This may, for example, be done by measuring carbonation depths, chloride penetration depths and thickness of reinforcement cover, as well as the w/c-ratio of the building component concerned. It may be the type of paint to be applied to achieve sufficient protection against penetration of carbon dioxide, CO_2 or chloride Cl^-.

Based on the specifications from the supplier/manufacturer of the equivalent concrete layer of the different surface protections, a list of possible surface protections is made. Together with any other requirements it is possible to determine the surface protection. Such other requirements may be:

- Surface protection on a downward concrete surface (e.g. a balcony) should always be more permeable to water vapour than an impermeable surface protection on upward surfaces. This is to prevent penetration of water through unexpected cracks or other defects in the upward surface protection which may lead to accumulation of penetrating water in the concrete (ponding) which again may lead to damage, such as moisture damage and frost damage to the concrete and blisters in the paint (membrane).
- The lifetime of surface protection on an upward concrete surface should exceed that of a downward concrete surface. The surface protection is decomposed with time (abrasion, aging, UV-radiation, etc.) and accumulation of moisture in the concrete should be prevented, see the above text.

Protection against water penetration

Membranes of acrylics, epoxy and polyurethane are impermeable and thus able to stop penetration of water. However, one situation should give rise to considerations, namely attempts to stop water penetration with a membrane on the inside of a concrete structure. In this situation the membrane will be subjected to pulling-off. Although this tensile action is less than the pull-off strength of the membrane, problems may arise. The membrane is affected by alkaline pore water and, for polyurethane membranes, in the form of slip and delamination. The phenomenon has been experienced on concrete containers which are partly below the ground water table and which have a polyurethane membrane on the inside.

Protection against penetration of aggressive substances through cracks

Surface protection of cracked concrete against penetration of aggressive substances requires knowledge of crack width changes versus time. It is easy to repair passive cracks (i.e. where the cause of cracking no longer exists), but repairing active cracks (where the cause of cracking still exists) and latent cracks (i.e. where the cause of cracking may rise again) requires considerable insight into the reaction of the concrete and surface protection.

By measuring the crack width versus time over an adequate period, it is possible to determine the flexibility which is a necessary property of a surface protection to protect against penetration of aggressive substances for a sufficiently long period. It is not only the expansion perpendicular to the crack, but also the shear between the two concrete areas on either side of the crack which may cause the paint film to degrade (i.e. the crack penetrates). It may be a difficult task which among others can be seen from the great number of (active) cracks that penetrate paint on concrete surfaces.

Migration of moisture and alkali reaction

Alkali reaction (in Denmark) is a chemical–physical reaction between porous flint particles in concrete aggregates and alkali from the cement of the putty or (rarely) alkali added from the outside. By the chemical reaction a reaction product (alkali–silica gel) is formed which can absorb moisture and expand so that cracks may develop in the concrete.

However, all conditions should be fulfilled in order that the alkali reaction can take place. Roughly speaking, the possibility of damage (cracking) together with the following conditions should be present:

- Concrete should contain a certain (high) quantity of alkali (dependent on the type of alkali-reactive aggregate), i.e. the concrete should be cast using average or high-alkali cement.
- The concrete aggregate should contain a certain quantity (over 2% by volume) of porous flint in the sand. If the sand fraction does not contain porous flint, some coarse grains of aggregate of porous flint in the coarse aggregate fraction may create local cracking (e.g. pop-outs).
- The concrete should have suitable moisture content, i.e. a relative humidity above 75% RH.

Since 1987 in Denmark, the use of concrete with very low probability of alkali reaction has been preferred. This is done by using aggregates without porous flint, low-alkali cement and a certain content of microsilica or fly ash to counteract formation of alkali reaction.

The moisture content of indoor direct-laid concrete floors (in passive exposure class) is below 75% RH. Therefore, the Danish concrete code does not require application of aggregates free of porous flint in passive exposure class or that high-alkali cement is not used.

Therefore, if an indoor direct-laid floor of (dry) concrete is rehabilitated so that the concrete becomes humid (i.e. above 75% RH) after rehabilitation, an alkali reaction may arise. Before design of rehabilitation of an indoor direct-laid (dry) concrete floor, it should be verified that the concrete is able to tolerate a change of exposure class (i.e. increased humidity).

An example of typical rehabilitation, which may cause increased humidity, is application of a impermeable membrane to an indoor direct-laid floor. It is generally assumed that the moisture content of a direct-laid concrete floor without surface protection (which is impermeable to water vapour) and with an adequate capillary-breaking layer is below 75% RH. If a membrane (of acrylics, epoxy or polyurethane), which is impermeable to water vapour diffusion, is applied, the moisture content of the concrete increases to above 90% RH.

Therefore, if an indoor direct-laid floor is made from concrete containing porous flint aggregates in either the sand or the coarse aggregate fraction with high-alkali cement as binder, crack inducing alkali reaction will be possible when the moisture content of the concrete increases to above 90% RH.

Therefore, as part of the design of membranes on direct-laid concrete floors, the composition of the concrete should be investigated.

Contractor/supervisor: surface protection

Execution of surface protection of concrete surfaces should conform to certain minimum requirements to fulfil the requirements of EN 1504-10 on 'Site application of products and systems and quality control of the works'. This applies to documentation of the repair materials received, the workmanship and the finishing treatment.

Requirements for workmanship

Surface protection of concrete surfaces should be performed so that:

- The specified repair materials are used
- Cleaning of the concrete surface does not damage the substrate
- The specified cleaning of the concrete surface is made
- The specified finishing treatment of the surface protection is made, for example, protection from the sun, desiccation or rain.

Reception control

EN 1504-2 on 'Surface protection systems for concrete' specifies a number of properties of surface protections which should be fulfilled with certain tolerances on reception, if identification testing (Table E1.9), is made. In order to be able to use Table E1.9 for identification testing, the supplier/manufacturer's data sheets should specify the declared expected values of properties and characteristics (in addition to the characteristic values for design purposes). Table E1.9 specifies the acceptable deviation from the expected values given by the supplier/manufacturer.

If the repair materials and systems bear the CE mark, identification testing can be replaced by checking the labels and branding of the materials concerned.

Table E1.9   Control by identification tests (deviations from specified values)

Properties and characteristics	Test method	Tolerance
Identification of components		
Colour and general appearance	Visual	Uniform and similar to the description on data sheet
Density	ISO 2811-1 or 2	Specified value ±3%
Infrared spectrum	EN 1767	a
Epoxy equivalent	EN 1877-1	Specified value ±5%
Amine equivalent	EN 1877-2	Specified value ±6%
Hydroxyl value	EN 1240	Specified value ±10%
Isocyanate content	EN 1242	Specified value ±10%
Volatile and non-volatile matter	EN ISO 3251	Specified value ±5%
Ash content	EN ISO 3451-1	Specified value ±5%
Thermogravimetry	EN ISO 11358	b
Flow time	EN ISO 2431	Specified value ±15%
Viscosity	EN ISO 3219	Specified value ±20%
Particle size distribution of dry components	EN 12192-1	>2 mm: ±6% abs. 0.063–2 mm: ±4% abs. <0.063 mm: ±2% abs.
Identification of fresh paint		
Glass beads method	EN ISO 1517	Specified value ±10%
Pot life (to 40 °C end temperature)	EN ISO 9514	Specified value ±15%
Hardness after 1, 3 and 7 days	EN ISO 868	±3 shore A or D after 7 days
Consistency	EN 1015-3	Specified value ±15% or ±20 mm
Air content	EN 1015-7	Specified value ±2% abs.
Bulk density	EN 12190 and EN 1015-6	Specified value ±5%
Workability – flow of mortar	EN 13395-2	Specified value ±15%
Stiffening time	EN 13294	Specified value ±20%

[a] The positions and relative intensities of the main absorption bands should match those of the reference spectrum.
[b] Confirmed by comparison, and ±5% with respect to loss of mass at 600 °C.

Preparation of substrate

The order of succession for surface treatment of concrete is:

- Cleaning
- Levelling
- Priming
- Painting.

Repair

If the concrete surface is to be repaired prior to surface treatment, the repair should comply with the requirements in EN 1504-3 on 'Structural and non structural repair', see Annex E6. Prior to surface treatment the repair should be hardened and the moisture content should be balanced with the level required by the supplier before surface treatment is commenced. Control of the moisture content of a repair may, for example, be made by a GANN meter.

Cleaning

The concrete surface is cleaned in order to open 'negative' blisters and make other defects visible so that voids, defects and depressions can be filled by the subsequent levelling. Furthermore, the cement skin on the concrete surface is removed so that the sand particles of the putty are exposed. This should be performed by abrasion with a carborundum or diamond wheel. Dry sand blasting may also be used, but this will cause dust nuisance. In case of sand blasting with added water the concrete surface is moistened. Sand blasting with added water is normally not recommendable, since a longer drying period is required for the concrete surface to fulfil the requirements of the supplier for the moisture of the concrete surface before application of primer and surface treatment (impregnation, sealing or paint).

In case of blast cleaning with air, it should be verified that the air is not polluted with oil or other substance which will reduce the adhesion ability.

For cleaning of large, horizontal upward concrete surfaces, shot blasting can be used. Shot blasting is machine cleaning of concrete surfaces where the cleaning agent (usually steel balls) is blasted towards the concrete surface and collected again in the machine. After filtering, the steel balls are re-used. The method is efficient for large areas, e.g. a floor. Shot blasting is dustless. Only a small layer of the concrete surface, typically only the casting skin, is removed. A shot cleaned concrete surface should always be checked and should be cleaned to remove loose particles from the concrete surface.

Regardless of the method of cleaning, the cleaned concrete surfaces should be solid, free of former surface protection, casting skin, dirt and visible dust that may rub off. The concrete surface should not be delaminated. The pull-off strength according to EN 1542 on 'Products and systems for the protection and repair of concrete structures. Test methods. Measurement of bond strength by pull-off' or EN 12636 on 'Determination of adhesion concrete to concrete' (pull-off testing on site) of uncracked concrete should at least be, see DS 411:1999, 3.2.4 (3)P or table 3.2.4:

- 1.1 MPa in passive exposure class: corresponding to compressive strength $f_c = 12$ MPa
- 1.6 MPa in moderate exposure class: corresponding to compressive strength $f_c = 25$ MPa
- 1.9 MPa in aggressive exposure class: corresponding to compressive strength $f_c = 35$ MPa
- 2.0 MPa in extra aggressive exposure class: corresponding to compressive strength $f_c = 40$ MPa.

If the pull-off strength is significantly lower than specified and failure occurs in the concrete it may indicate that the concrete surface may be delaminated. Therefore, a further investigation of the concrete surface will be expedient, e.g. by 'impact echo'.

The mode of failure for pull-off testing of cleaned concrete surfaces should be failure in the concrete. Surface protection of very weak concrete surfaces requires use of special techniques, e.g. in-depth impregnation.

Levelling

The cleaned concrete surface should be levelled to achieve a uniform appearance and an efficient protection of the concrete surface, since voids, defects and cracks will be filled with the levelling product. It is a precondition for achieving an efficient protection from penetration of aggressive substances.

It is important that cracks are filled. A special blister formation in membranes may occur if cracks in the concrete surface have not been filled prior to priming and application of a membrane. This phenomenon may occur in membranes in, for example, bridge decks and balconies under special circumstances. It may be explained by the following example:

A membrane should be applied to an outdoor concrete deck. After cleaning of the concrete slab a primer and then an epoxy membrane are applied. There is some cracking (crazing) in the surface. The crack width is approximately 0.1 mm and is considered insignificant to the membrane application.

The chosen primer is highly fluent with a long curing time to ensure proper penetration into the concrete surface prior to application of the epoxy membrane. Since the primer is highly fluent it is not particularly crack-bridging when wet. Because of the long curing time of the primer (approximately 10 hours), major temperature fluctuations will occur during the curing period. The temperature of the concrete will rise from the start of the working day in the morning and throughout the day. At the end of the working day the temperature of the concrete will drop at nightfall and rise again at sunrise. The cracks are filled with air which will expand when the temperature rises. If the primer has not cured properly so that it is able to sustain the air pressure, a blister will form above the crack. If this is not detected and an epoxy membrane is applied, this phenomenon may repeat itself leading to blisters in the epoxy membrane.

Such blistering may be avoided by using filling (levelling) or a fast-curing primer with suitable crack-bridging ability when wet. However, staggered working hours (work shifts) may be applied so that the primer is applied during a period of dropping temperature (in the concrete) and so that the air in the cracks shrinks. Furthermore, the curing time should be adapted so that the primer has obtained adequate curing when the temperature starts rising at sunrise.

The climate should not be harmful to the levelling so that, for example, plastic cracks are formed. Normally it may be considered acceptable if the climate fulfils the following conditions:

- The temperature of the concrete surface should not be below 5 °C at the coldest time of the day during the curing period of the levelling (the lowest temperature given in the data sheet)
- The levelling should not be exposed to driving rain
- The levelling should not be exposed to direct sun or strong winds.

Priming

Certain paints and membranes for protection of concrete surfaces require priming. Normally specific primers apply to each paint and membrane for the surface protection of concrete. Primers are applied to make a sustainable, cleaned and levelled concrete surface suitable for receiving the actual surface protection, i.e.:

- An absorbent base should be made non-absorbent by priming
- Priming should form a barrier in the concrete to prevent salts from penetrating the surface protection and lead to discolouration (especially sensitive in dark paint)
- Lightly rubbing-off bases should be made sustainable by priming.

The climate should not be harmful to the priming, for example, the primer being washed-off by driving rain. Normally it is considered acceptable if the climate fulfils the following conditions:

- The temperature of the concrete surface should not be below 5 °C at the coldest time of the day during the curing period of the levelling (the lowest temperature given in the data sheet)

- The ambient humidity should not exceed 80% RH at the most humid time of the day during the curing period of the primer (maximum humidity should be specified on the supplier's data sheet).

Priming is a process which is significant for the durability, especially of membranes. Thus, blistering may occur in a membrane of acrylics, epoxy or polyurethane applied to a concrete surface for the following reasons:

- The component applied is a dual component. If the mix proportion is not correct, blistering may occur. Thus there may be 'saponification' in 'pockets' with incomplete polymerization. Generally this phenomenon is followed by an acid smell (as of vinegar) of the liquid of punctured blisters.
- Blistering may occur if the membrane is applied to a too moist concrete surface. This blistering (due to vapour) will occur shortly after the application.
- Blistering may occur due to formation of osmosis cells. Blistering due to osmosis often occurs in cases where the primer applied is too thin and perhaps is missing in some areas. Thus semipermeable areas are created, e.g. at 'pinholes'. An osmosis cell may be formed which leads to a dent in the membrane with a diameter of approximately 20–30 mm. Segregation of the entire membrane is, however, hardly possible because concrete is normally too permeable so that an osmotic pressure cannot develop and be kept up.

Paint and membranes

Finally, the paint prescribed in the design is applied by the specified tool. The paint is usually applied in several coatings. The specified paint film thickness should be measured and documented. By impregnation and sealing (e.g. concrete sealer) no paint film is formed. Instead, documentation of material consumption for especially chosen inspection sections will be required.

The climate should not be harmful to the priming, e.g. primer being washed-off by driving rain. Normally it is considered acceptable if the climate fulfils the following conditions:

- The temperature of the concrete surface should not be below 5 °C at the coldest time of the day during the curing period of the levelling (the lowest temperature given in the data sheet).
- The ambient humidity should not exceed 80% RH at the most humid time of the day during the curing period of the primer (maximum humidity should be specified on the supplier's data sheet).
- The paint should not be exposed to direct sun or strong winds.

Control of the works

In Tables 6.2.1, 6.2.2 and 6.3.1 of EN 1504-10 on 'Site application of products and systems and quality control of the works' it is specified how quality control of the application of sealing, paint and impregnation should be performed.

Pull-off testing

Control of concrete surfaces which are rehabilitated by application of a wearing course or membrane is usually performed by pull-off tests. In this way the cohesive strength of the concrete or its adhesive strength on the substrate is determined. Previously, no test standard for pull-off tests existed in Denmark. Therefore, a pull-off test of concrete surfaces was performed according to the test standards of other countries. The result was that there was no guarantee that two tests (according to different standards) were performed under identical test conditions and therefore comparable.

This has now been changed since EN 1542:1999 on 'Products and systems for the protection and repair of concrete structures. Test methods. Measurement of bond strength by pull-off' is now assumed to apply in connection with the EN 1504 standards. EN 1542 is a laboratory test which presupposes that test specimens of the dimensions $300 \times 300 \times 100$ mm can be cast, rehabilitated and controlled by pull-off testing. However, in Denmark such pull-off testing is traditionally performed on the concrete structure itself. Thus, local conditions such as casting flaws in the substrate and incomplete curing are taken into consideration. The same applies to the standard EN 1015-12 on 'Methods of test for mortar for masonry – Part 12: Determination of adhesive strength of hardened rendering mortars on substrates'.

The advantage of EN 12636 is that pull-off testing may now be identical from one concrete structure to another and independent of the test standard chosen for the control task concerned.

Pull-off testing (on a test specimen or in a structure) should be distinguished in the following typical cases:

- cleaned concrete surface of substrate
- placed and cleaned mortar layer on substrate
- applied and cleaned membrane on substrate (e.g. mortar layer).

Test procedure

Control of concrete surfaces protected or rehabilitated by paint and membranes may be performed by pull-off testing. The test methods are EN 1542 on 'Products and systems for the protection and repair of concrete structures – Test methods – Measurement of bond strength by pull-off'. The methods in EN 1542 may also be used for site measurement on layers exceeding 0.5 mm. For thinner layers, however, the cross cut test method as described in EN ISO 2409-6 should be used.

Control of paint film thickness

In addition to control of the pull-off strength of the surface protection, control of the paint film thickness is also important. The methods used are:

- *Calculated paint film thickness*. The average paint film thickness can be calculated by the method described in ISO 7254:1986. In order for the calculated layer thickness to be usable it is a condition that the paint has been evenly applied. The calculated layer thickness of the paint film is the average thickness over the entire concrete area concerned.
- *Non-destructive measurement*. Prior to application of the surface protection, small pieces of metallic foil are glued on (magnetic and non-magnetic metals) to the concrete area. After application of the surface protection the layer thickness can be measured by a thickness meter, e.g. List-Magnetics μ-meter. The presence of the pieces of metallic foil may, however, influence the personnel performing the surface protection.
- *Destructive measurement*. If a cut through the paint layer can be made it is possible to measure the layer thickness of the dry paint film by microscope, or the thickness of the paint layer can be measured on, for example, a bored core. According to ISO 2808:1991

the sample is embedded in (dyed) epoxy and the cut forming an angle to the surface is abraded. Then the film thickness is measured by microscope and specified in μm, corresponding to the layer thickness of dry paint film.

References

Debaiky, Green, Hope. Corrosion of FRP-wrapped RC cylinders – long term study under severe environmental exposure. Proceedings of the FRPRCS-5 Vol. 2. University of Cambridge. Thomas Telford, UK, 2001.

Jensen. Relative humidity measured by the wooden stick method in Norwegian concrete structures with and without surface protection. Proceedings Nordic Concrete Research Meeting. Elsinore, Denmark, 2002.

Justeness, Søpler. The performance of polymer impregnated concrete beams after 19 years of outdoor exposure. Nordic Concrete Research no. 11. Oslo, Norway, 1992.

Søpler. Concrete polymer materials. FCB report no. 2. Trondheim, Norway, 1971.

Vaysburd, McDonald. An evaluation of equipment and procedures for tensile bond testing of concrete repairs. US Corps of Engineers, Waterways Experiment Station. Washington, USA, 1999.

Wehner. Arbeitsanweisung für kombinierte Griffigkeits- und Rauheitsmessungen mit dem Pendelgerät und dem Ausflussmeber. Forschungsgeshellschaft für das Strassenwesen, Arbeitsgruppe: Fahrzeug und Fahrbahn. Anhang: Zur Problematik der kombinierten Griffigkeits- und Rauheitsmessungen mit dem Pendelgerät und dem Ausflussmesser. Köln, Ausgabe, 1972.

Annex E2: Injection of cracks, defects and voids

Introduction

The standard EN 1504-5 on 'Concrete injection' describes injection of cracks, defects and voids in concrete building components where the purpose of the injection is:

- Principle P1. Sealing of the concrete
- Principle P4. Strengthening of the concrete.

The purpose of injection of cracks in concrete is to recreate original monolithic properties and characteristics of the building component, i.e. adequate strength, stiffness and density. Thus distinction between sealing and filling of cracks shall be made to: make the building component impermeable to penetration of water or aggressive substances without strengthening the component (method M1.4); make the building component impermeable to water or aggressive substances and strengthening the concrete (methods M4.5 and M4.6).

Requirements for injection materials are different depending on their intended uses. Requirements for injection materials and their application are made in EN 1504-5 on 'Concrete injection'.

Background for methods M1.4, M4.5 and M4.6

Concrete in damaged building components is often cracked. Dependent on the circumstances, injection of cracks may be chosen, i.e. use of a special technique by which the crack is filled (injected) with repair material suitable for the purpose.

Cracks

Normally reinforced concrete has cracks, and crack widths (on the concrete surface) will normally be acceptable if they are within the maximum limit (i.e. the largest crack width allowed) in the Eurocode ENV 1992-1-1 and EN 206-1 or concrete codes valid in current use. However, cracks may be formed due to actions other than those from a stress field (bending, shear and torsion). Cracks may be caused by, for example, shrinkage, thermal stress, corrosion and overload. Not all cracks look alike, neither in general (pattern) nor in detail (width and depth). Thus cracks may be classified into the following categories:

- Through-going cracks, i.e. cracks which penetrate through a building component and which have at least two different concrete surfaces
- Local cracks, i.e. cracks where the propagation is concentrated locally in a building component
- Widely distributed cracks, i.e. cracks spread over the entire structural member with little variation of intensity.

Changes of crack width are of importance for the choice of rehabilitation technique. Therefore, it is practical to divide cracks into the following categories:

- *passive* ('dead') cracks, i.e. the crack is not moving and the cause of cracking is no longer there, i.e. old shrinkage cracks
- *active* ('live') cracks, i.e. the crack is still moving and the crack width changes. The cause of cracking still exists, e.g. durability cracks (alkali reaction, rust) and cracks due to fluctuating thermal action (temperature)
- *latent* ('dormant') cracks, i.e. cracks which appear to be passive but may become active by rehabilitation.

Concrete with cracks where the crack width on the concrete surface is $w \geqslant 0.2$ mm can normally be filled by injection and a suitable injection material.

Injection of cracks

It is necessary to distinguish between:

- Pressurized injection where the injection fluid is pressed into the crack through nipples bored into the concrete surface
- Surface injection where part of the concrete is subjected to vacuum with subsequent pressurized injection. This method is not included in EN 1504-5 on 'Concrete injection'
- Crack filling or gravitational injection where the crack is filled with filling material by gravitation.

Injection of defects and interstices

While cracks are formed after casting of the structure, defects and interstices will be created during pouring of the concrete. Normally it is possible to fill or inject such defects and interstices and the negative effects can thereby be eliminated.

Injection materials

EN 1504-5 includes the following groups of injection materials:

- Force transmitting filling of cracks, defects and interstices with repair materials able to bond to concrete surfaces in cracks, defects and interstices so that transmission of force is possible. Strict requirements are made for the failure criterion, failure deformation and bonding of the repair materials in the actual temperature and moisture intervals. Often injection materials will be brittle after hardening. Force transmitting injection is typically made with epoxy.

- Injection materials which are plastic after hardening so that cracks, defects and interstices after filling are able to absorb a subsequent (minor) movement. Typical injection materials are polyurethanes.
- Injection materials which swell in connection with hardening and water absorption so that cracks, defects and interstices are filled very efficiently. In this class of injection materials are certain gels which, however, only have sealing effect when in constant contact with water. Normally these gels do not bond very well to the substrate. Typically, a swelling injection material is acrylic gel.

EN 1504-5 on 'Concrete injection' distinguishes between two main types of injection materials, namely:

- Polymeric injection materials where the hardening is related to a reactive polymer binder, e.g. epoxy, polyurethane and acrylics. Polymeric injection materials are able to stop moisture migration in the concrete, for example.
- Hydraulic injection material where the hardening is related to hydration reaction of a hydraulic binder, e.g. cement paste of ultra-fine cement. Hydraulic injection materials allow moisture migration in concrete.

Classification of injection products

The manufacturer/supplier should classify injection materials according to the performance requirements. EN 1504-5 applies the following classification, e.g. U(F) W(1)(1/2)(5/30) D(1)(0), where:

- U identifies the group followed by a parenthesis where:
  - F signifies force transmitting injection material
  - D signifies ductile injection material
  - S signifies swelling injection material
  - W signifies conditions for crack and environment followed by three parentheses where: allowed minimum thickness of crack, measured in tenths of millimetres (0.1, 0.2, 0.3, 0.5 and 0.8 mm); allowed moisture state at which cracks in concrete are injectable with the injection material concerned (1 for dry, 2 for damp, 3 for wet and 4 for water flowing); minimum and maximum use temperature of the injection.
- D denotes durability followed by two parentheses:
  - In the first parenthesis 1 is stated if the bonding is durable at a water-bearing crack and 0 if not (or not documented)
  - In the second parenthesis 1 is stated if the injection material is compatible with elastomers and 0 if not (or not documented).

An injection material, which is, for example, classified as U(F) W(1)(1/2)(5/30) D(1)(0), signifies that:

- the product is a force transmitting injection material
- the product is usable for injecting cracks with a crack width of 0.1 mm
- the product is usable for injecting dry as well as damp cracks
- the product is usable at use temperatures between 5 °C and 30 °C
- the bonding of the product is durable when wet
- the compatibility of the product with elastomers is not documented.

Table E2.1 Performance of force transmitting injection materials according to the methods M4.5: 'Injection of cracks, voids and interstices' and M4.6: 'Filling of cracks and interstices'

Properties of force transmitting injection materials according to EN 1504-5	Test method	Polymeric injection materials	Hydraulic injection materials
Basic properties and characteristics			
Adhesion by tensile bond strength	EN 12618-2	■	■
Adhesion by slant shear strength	EN 12618-3	▲	▲
Volumetric shrinkage	EN 12617-2	■	~
Water segregation (bleeding)	EN 445/3.3	~	■
Volume change	EN 445/3.4	~	■
Glass transition temperature	EN 12614	▲	~
Chloride content	EN 196-21	~	▲
Workability			
Injectability into dry medium[a]	EN 1771	■	■
Injectability into non-dry medium[a]	EN 1771	■	■
Viscosity	EN 3219/EN 14117	■	■
Curing conditions			
Workable life	EN ISO 9514	■	■
Tensile strength development	EN 1543	■	~
Setting time	EN 196-3	~	■
Durability			
Adhesion by tensile bond strength after thermal and wet-drying cycles	EN 12618-2	■	■
Compatibility with concrete	EN 12618-2	■	■

■, for all intended uses; ▲, for certain intended uses; ~, no requirements.
[a] Where EN 1771 does not apply and in cases where the injectability classes 0.5 mm and 0.8 mm should be used, EN 12618-2 should be applied.

EN 1504-5 has classified the three types of injection materials into the following groups:

- F: force transmitting injection material
  - F1: bonding for pull-off strength above 2 MPa for injection (Table E2.1)
  - F2: bonding for pull-off strength above 0.6 MPa for filling of cracks (Table E2.1)
  - D: ductile injection material
  - D1: impermeable at 2×10^5 Pa (Table E2.2)
  - D2: impermeable at 7×10^5 Pa for special uses (Table E2.2)
- S: swelling injection material
- S1: impermeable at 2×10^5 Pa (Table E2.3)
- S2: impermeable at 7×10^5 Pa for special uses (Table E2.3).

Materials for injection

Injection materials should possess certain properties to be suitable for injection of cracks either to achieve density or density as well as strength (Tables E2.1–E2.3). The standard EN 1504-5 on 'Concrete injection' requires that injection materials should have certain properties all of which should be specified in order that the products concerned may be

Table E2.2   Performance description for ductile filling of cracks according to the methods M1.4: 'Filling of cracks, voids and interstices'

Properties of ductile injection materials according to EN 1504-5	Test method	Polymeric injection materials	Hydraulic injection materials
Basic properties and characteristics			
Adhesion to concrete and deformability	EN 12618-1	■	~
Watertightness	EN 14068	▲	~
Glass transition temperature	EN 12614	▲	~
Workability			
Injectability into dry medium[a]	EN 1771	■	~
Injectability into non-dry medium[a]	EN 1771	■	~
Viscosity	EN ISO 3219	■	~
Expansion rate	EN 14406	▲	~
Curing conditions			
Workable life	EN ISO 9514	■	~
Durability			
Compatibility with concrete	EN 12637-1	□	~

■, for all intended uses; ▲, for certain intended uses; □, for special intended uses; ~, no requirements.
[a] Where EN 1771 does not apply and in cases where the injectability classes 0.5 mm and 0.8 mm should be used, EN 12618-2 should be applied.

chosen for injection of cracks, defects and interstices in concrete. Furthermore, minimum requirements are made for these properties (Tables E2.4–E2.6).

Requirements for documentation

The manufacturer or supplier of injection materials should be able to document a number of properties and characteristics (data sheets) of the marketed injection materials. Table E2.1 shows an overview of force transmitting injection materials which should be documented for all intended uses, for certain intended uses and for special purposes. Table E2.2 applies to ductile injection materials and Table E2.3 applies to swelling injection materials.

Minimum requirements

Properties and characteristics should be documented in data sheets for injection materials. EN 1504-5 on 'Concrete injection' prescribes the minimum requirements to be fulfilled for injection materials. Minimum requirements for specific properties and characteristics and required test methods can be seen from Table E2.4 for force transmitting injection materials, Table E2.5 for ductile injection materials and Table E2.6 for swelling injection materials.

Designer/builder: design

A rehabilitation project should fulfil certain property requirements for injection of cracks, defects and interstices in a building component. EN 1504-5 on 'Concrete injection' lists

Table E2.3 Performance of swelling injection materials according to method M1.4: 'Filling of cracks, voids and interstices'

Properties of swelling injection materials according to EN 1504-5	Test method	Polymeric injection materials	Hydraulic injection materials
Basic properties and characteristics			
Watertightness[a]	EN 14068	■	~
Rust preventing capacity[b]	~	▲	~
Workability			
Working life and viscosity[c]	EN ISO 3219	■	~
Expansion ratio and rate by water storage[d]	EN 14498	■	~
Curing conditions			
Pot life[e]	EN 9514	■	~
Durability			
Sensitivity to water[f]	EN 14498	■	~
Sensitivity to wet-drying cycles[g]	EN 14498	■	~
Compatibility with concrete[h]	EN 12637-1	■	~

■, for all intended uses; ▲, for certain intended uses; ~, no requirements.

[a]The test method EN 14068 should be supplemented with 500 compressive cycles. After exposure to the declared maximum compression for 7 days as specified in EN 14068, the compression should be reduced to 50% of the declared maximum compression and be maintained for 2 hours prior to the cyclic variations of the compression. Each of the mentioned 500 compressive cycles should be maintained for 15 minutes at 75% of the maximum compression followed by 15 minutes at 25% of the maximum compression.

[b]Until an accepted EN standard for the rust protection capacity of injection materials is available, the national rules should be observed when required.

[c]In cases where EN ISO 3219 does not apply, EN 12618-2 should be used. For crack widths above 0.3, 0.5 and 0.8 mm spacers of imperishable plastics should be applied to fix the crack width used.

[d]Expansion due to moisture is measured by changes of volume and mass by air-drying and water storage.

[e]Testing should be performed at three test conditions: At 21 °C and the minimum and maximum temperature recommended by the manufacturer. The test temperature should be kept within a tolerance of ±1 °C.

[f]Expansion due to water absorption is measured by changes of volume and mass by air curing and water storage.

[g]Sensitivity to wet-drying cycles is measured by changes of volume and mass at cycling air-drying and water storage.

[h]The test should be performed according to EN 12637-1, applying six 15 mm thick test specimens. Three test specimens are stored in tap water and three are stored in a 1 M KOH solution.

minimum requirements for these properties, all of which should be specified and documented for use as injection materials for concrete. Furthermore, test methods for these properties are stated (Tables E2.4–E2.6).

Requirements for design of force transmitting injection

Static calculation of force transmitting, injected cracks should be performed in accordance with Eurocode ENV 1992-1-1 and EN 206-1 or concrete codes valid in current use. Here it is specified that durability requirements for the injected building component should be fulfilled. It should be noted, however, that the requirements for injection materials are not necessarily adequate for special environments and special loads, e.g. cryogen, (extremely low) temperature, impacts from traffic, ice and earthquakes.

Table E2.4   Minimum requirements for force transmitting injection materials and crack fillers

Properties and characteristics	Test method	Polymer injection materials	Hydraulic injection materials
Properties and characteristics			
Pull-off strength	EN 12618-2	Failure in concrete	>2 MPa >0.6 MPa for injection products intended for filling cracks
Adhesion by slant shear strength[a]	EN 12618-3	Monolithic failure	Monolithic failure
Volumetric shrinkage	EN 12617-2	<3%	~
Water segregation (bleeding)	EN 445/3.3	~	<1% after 3 hours
Volume change	EN 445/3.4	~	<5% shrinkage <1% expansion
Glass transition temperature	EN 12614	>40 °C	~
Chloride content[b]	EN 196-21	~	<0.2% concrete
Workability			
Injectability into dry medium and splitting strength f_s for crack widths 0.1 mm, 0.2 mm and 0.3 mm[c]	EN 1771	Injectability class: 0.1: <4 min 0.2 and 0.3: <8 min f_s > 7 MPa	Injectability class: 0.1: <4 min 0.2 and 0.3: <8 min f_s > 3 MPa
Injectability into non-dry medium and splitting strength f_s for crack widths 0.1 mm, 0.2 mm and 0.3 mm[c]	EN 1771	Injectability class: 0.1: <4 min 0.2 and 0.3: <8 min Splitting strength >7 MPa	Injectability class: 0.1: <4 min 0.2 and 0.3: <8 min Splitting strength >3 MPa
Viscosity	EN ISO 3219 and EN 14117	Declared value	Declared value
Curing conditions			
Pot life[d]	EN 9514	Declared value	Declared value
Strength development	EN 1543	Declared value	Declared value
Setting time	EN 196-3	Declared value	Declared value
Durability			
Adhesion by tensile bond strength after thermal and wet-drying cycles	EN 12618-2	100% failure in concrete	Reduction <30%
Compatibility with concrete	EN 12618-2	100% failure in concrete	Reduction <30%

[a] Monolithic failure is in this case defined as the situation where an injected test specimen fails with the same mode of failure as a solid-cast test specimen.

[b] The requirements for maximum chloride content are specified with a too high limit (according to Danish traditions). There is no reason to relax these requirements, since it is not difficult to fulfil a less strict requirement. Requirements for chloride content of maximum 0.02–0.05% of the mass of the injection material should be easily fulfilled.

[c] In cases where EN 1771 does not apply and where the injectability classes 0.5 and 0.8 mm are relevant, EN 12618-2 can be used. For the injectability classes 0.3–0.5 and 0.8 mm, inert flexible plastic spacers of, respectively, 0.3–0.5 and 0.8 mm should be used.

[d] The test should be in accordance with EN 9514, but a quantity of 1000 ml instead of 300 ml should be used and three conditioning and test temperatures should be used: 21 ± 2 °C. The minimum use temperature recommended by the manufacturer/supplier with a tolerance of ±1 °C. The maximum use temperature recommended by the manufacturer/supplier with a tolerance of ±1 °C. Pot life is defined as the period of time from the mix has been finished until the temperature of the mix has risen 15 °C for polymeric injection materials; the filtration stability is terminated by the value declared by the manufacturer/supplier for hydraulic injection materials.

Table E2.5 Minimum requirements for plastic (polymeric) injection materials

Properties of plastic injection materials	Test method	Polymeric injection materials
Properties and characteristics		
Adhesion to concrete and plasticity	EN 12618-2	Adhesion: Declared value Elongation: >10%
Watertightness	EN 14068	Approximately 2×10^5 Pa and approximately 7×10^5 Pa for special intended uses
Glass transition temperature	EN 12614	For information
Workability		
Injectability into dry medium for crack widths of 0.1 mm, 0.2 mm and 0.3 mm[a]	EN 1771	Injectability class: 0.1: <4 min Injectability class: 0.2 and 0.3: <8 min
Injectability into non-dry medium[a]	EN 1771	Injectability class: 0.1: <4 min Injectability class: 0.2 and 0.3: <8 min
Viscosity	ISO 3219	Declared value
Expansion rate	EN 14406	Declared value
Curing conditions		
Pot life[b]	EN 9514	For information
Durability		
Compatibility with concrete	EN 12637-1	No failure by compressive testing Lost deformation work <20%

[a] In cases where EN 1771 does not apply and where the injectability classes 0.5 and 0.8 mm are relevant, EN 12618-2 can be used. For the injectability classes 0.3–0.5 and 0.8 mm, inert flexible plastic spacers of, respectively, 0.3–0.5 and 0.8 mm should be used.

[b] The test should be in accordance with EN 9514, but a quantity of 1000 ml instead of 300 ml should be used and three conditioning and test temperatures should be used: 21 ± 2 °C. The minimum use temperature recommended by the manufacturer/supplier with a tolerance of ±1 °C. The maximum use temperature recommended by the manufacturer/supplier with a tolerance of ±1 °C. Pot life is defined as the period of time from the mix has been finished until the temperature of the mix has risen 15 °C for polymeric injection materials, and the filtration stability is terminated by the value declared by the manufacturer/supplier for hydraulic injection materials.

Requirements for filling of cracks

If a crack is not required to be force transmitting, but only impermeable to liquids, gases, staphylococci, radon, etc., no static calculation is required for the injected crack, but only documentation of impermeability.

Registration of environment and state

Before design of repair of cracks, they should be registered together with their local environment, including:

- *Crack width.* Five injectability classes are defined corresponding to the smallest crack width in mm measured on the surface of the substrate. The five classes are 0.1, 0.2, 0.3, 0.5 and 0.8 mm.

Table E2.6   Minimum requirements for swelling (polymer) injection materials according to EN 1504-5

Properties of swelling injection materials	Test method	Polymeric injection materials
Properties and characteristics		
Watertightness[a]	EN 14068	Approximately 2×10^5 Pa and approximately 7×10^5 Pa for special intended uses
Corrosion protection[b]	~	~
Workability		
Viscosity/workability	ISO 3219	For information
Expansion and water absorption	EN 14498	For information
Curing conditions		
Pot life[c]	EN 9514	For information
Durability		
Sensitive to water by water storage	EN 14498	The expansion ratio should reach a constant level during the water immersion
Sensitive to wet/drying cycles	EN 14498	After wet-drying cycles there should be no change to the expansion ratio after water immersion
Compatibility with concrete[d]	EN 12637-1	Limited reduction in strength

[a] The test method EN 14068 should be supplemented with 500 compressive cycles. After exposure to the declared maximum compression for 7 days as specified in EN 14068, the compression should be reduced to 50% of the declared maximum compression and be maintained for 2 hours prior to the cyclic variations of the compression. Each of the mentioned 500 compressive cycles should be maintained for 15 minutes at 75% of the maximum compression followed by 15 minutes at 25% of the maximum compression.

[b] Until an accepted EN standard for the rust protection capacity of injection materials is available, the national rules should be observed when required.

[c] The test should be in accordance with EN 9514, but a quantity of 1000 ml instead of 300 ml should be used and three conditioning and test temperatures should be used: 21 ± 2 °C. The minimum use temperature recommended by the manufacturer/supplier with a tolerance of ± 1 °C. The maximum use temperature recommended by the manufacturer/supplier with a tolerance of ± 1 °C. Pot life is defined as the period of time from the mix has been finished until the temperature of the mix has risen 15 °C for polymeric injection materials. The filtration stability is terminated by the value declared by the manufacturer/supplier for hydraulic injection materials.

[d] The test should be performed according to EN 12637-1, applying six 15 mm thick test specimens. Three test specimens are stored in tap water and three are stored in a 1 M KOH solution. After 24 hours of curing at standard conditions, the injection material (from the pot life test) is subjected to a concentrated compressive load Ø 20 mm with conical head (top angle 60°) at a rate of 100 mm/min. The load–deformation curve should be recorded. The strength properties should not be reduced more than 20% compared to water-stored test specimens, see EN 12639-1.

- *Moisture in cracks*. Four degrees of moisture for cracks are defined: dry, damp, wet and water flowing.
- *Activity of cracks*. Three types of activity are defined: active (live), passive (dead) and latent (dormant) cracks.

Other conditions may be registered, but the above are necessary to be able to choose the type of injection material.

Injection materials

EN 1504-5 on 'Concrete injection' distinguishes between injection materials with polymer binders and injection materials with hydraulic binders. Different requirements are made for the two types, and for each type there are products with different performances. Based on the designer's registration, the injection product or products (according to the supplier's data sheets), which fulfil the necessary performance requirements, are chosen.

Safety, health, environment and resistance to fire

ENV 1504-9 does not make specific requirements for safety, health, environment and resistance to fire in connection with products and systems for protection and repair of concrete structures, but requires that relevant measures are taken. This means that national requirements should be observed, i.e. the regulations (in Denmark) of The Danish Working Environment Service and fire regulations.

Contractor/supervisor: execution of work

Execution of injection of cracks, defects and interstices should conform to certain minimum requirements to fulfil the requirements in EN 1504-10 on 'Site application of products and systems and quality control of the works'. This applies to documentation of the injection materials received, execution of work and finishing treatment.

Requirements for execution of work

Injection of cracks, defects, etc., should be performed according to the following principles:

- Preparation of the substrate is made as specified in the section 'Method'
- Preparation of cracks is made as specified in the section 'Method'
- Injection is made as specified in the section 'Method'
- Specified injection materials should be used
- Injection of a crack should always be finished in one operation because re-injection is not possible, once the injection material has hardened
- An injected crack is not finished before it has been 're-injected', i.e. all nipples in a crack have been re-injected before the injection material in the crack has hardened.

Method

Filling of cracks, defects and interstices by injection can be performed by compression, gravitation or vacuum. The allowable temperature range for the substrate should be observed (only injection at temperatures above zero).

Prior to injection, impurities in cracks, defects and interstices should be removed if possible (e.g. by water or air). Aggressive liquids such as hydrochloric acid, HCl, should not be used for cleaning. The permissible moisture content in cracks, defects and cavities to be injected depends on the injection material. However, it is possible to inject against water pressure with special injection materials.

Cracks from corrosion on reinforcement due to chloride attack should not be injected before concrete with chloride content above the threshold value has been removed or the chloride content has been reduced, e.g. by electrochemical chloride extraction.

Choice of injection fluid

As previously mentioned, cracks which are down to 0.2 mm wide on the concrete surface can be injected with a suitable injection fluid. The less the crack width, the more fluent the injection fluid should be. A crack width in concrete above 3 mm requires a viscosity of 350–500 m·Pa·s. Very thin injection fluids can be used for very narrow cracks or for coarse cracks as pre-injection. The thin injection fluids are absorbed in the capillary pores. Then heavier fluids (350–500 m·Pa·s) are injected wet-on-wet.

Very coarse cracks (3 mm and above) require thixotropic injection fluid in order that the injection fluid will remain inside the crack. This can be achieved by adding filler to the injection fluid, e.g. quartz. The following information can be given of the individual injection products:

- Epoxy should have low viscosity since it changes significantly with temperature, i.e. becomes thin at elevated temperatures and thick at low temperatures. Epoxy should not normally be used at temperatures below 8 °C. Furthermore, it should be assumed that both pot life and curing time will also change with the viscosity at higher and lower temperatures, respectively.
- Polyurethane, two-component, i.e. types consisting of a resin (polyol) component and a curing component (isocyanate), which should also have low viscosity. Also the viscosity changes at higher and lower temperatures, but it is not as sensitive as epoxy. Two-component polyurethane should not normally be used at temperatures below 5 °C.
- Polyurethanes, one-component, normally foaming types (including polyurethanes where a catalyst or an accelerator should be added before they can perform). These are even less sensitive to higher or lower temperatures than the two-component polyurethanes. They can also be applied at temperatures below zero provided there is no ice in the cracks.
- Gels including acrylic gels (amide types and amine types) contain a large proportion of ordinary water. The gels are not particularly sensitive to temperature changes.
- Cement injection materials are not particularly sensitive to temperature changes, but they set faster at elevated temperatures and slower at lower temperatures.

Reception control

EN 1504-5 on 'Concrete injection' specifies a number of properties and characteristics of injection materials (components, fresh and hardened injection material), which should be fulfilled within certain tolerances on reception, see Table E2.7 for polymeric injection materials and Table E2.8 for hydraulic injection materials.

If, on the other hand, the injection materials and systems are CE-labelled, identification testing can be replaced by checking the tags and labels of the materials concerned.

Control of workmanship

Injection of cracks is specialist work. Control of workmanship can be performed by the supervisor monitoring the injection. Control of compliance with mixing time is particularly important. In case of doubt in the workmanship, unfilled cracks can be revealed by 'impact echo' and if necessary boring of a core where a thin section is analysed by UV radiation. Such an analysis is specialist work.

Table E2.7 Identification requirements for injection products formulated with reactive polymer binder

Basic properties and characteristics	Test method	Tolerance (% deviation from data sheet)
Individual components		
Epoxy equivalent	EN 1877-1	Declared value ±5%
Amine functions	EN 1877-2	Declared value ±6%
Hydroxyl value	EN 1240	Declared value ±10%
Isocyanate content	EN 1242	Declared value ±10%
Other functional groups[a]	Not specified	~
Specific weight	EN/ISO 2811-1	Declared value ±3%
Infrared analysis[b]	EN 1767	Corresponding to ref.
Identification of freshly mixed product		
Viscosity[c]	ISO 3219	Declared value ±20%
Pot life[d]	EN ISO 9514	Declared value ±20%
Volatile matter	EN ISO 3251	Declared value ±5%
Identification of the hardened mixture		
Tensile strength, elongation and elastic modulus[e]	EN ISO 527-1 and 2	Declared value ±20%
Strength properties[f]	~	Declared value ±20%

[a]Determination of other components may be significant. However, no requirement or test standard is specified.

[b]The measured positions and relative intensities of the absorption strips should correspond to the reference given by the manufacturer.

[c]Viscosity should be measured 5 minutes after mixing of the product has been completed. The temperature of the components should be maintained at 21 ± 2 °C before mixing. The temperature of the freshly mixed product should be measured and recorded. For injection materials which harden in less than 5 minutes, viscosity should be measured on unmixed components.

[d]The test should be in accordance with EN 9514, but a quantity of 1000 ml instead of 300 ml should be used and three conditioning and test temperatures should be used: 21 ± 2 °C. The minimum use temperature recommended by the manufacturer/supplier with a tolerance of ±1 °C. The maximum use temperature recommended by the manufacturer/supplier with a tolerance of ±1 °C. Pot life is defined as the period of time from when the mix was finished until: the temperature of the mix has risen 15 °C for polymeric injection materials, and the filtration stability is terminated by the value declared by the manufacturer/supplier for hydraulic injection materials.

[e]Only injection material for filling and other ductile injection materials should be tested according to EN 527-1. The test should be performed after 7 days of curing under standard conditions for polymeric injection materials. The test should be performed on a poured sheet (thickness 3 mm) of the injection material concerned.

[f]The following method should only be applied to swelling injection materials: After 24 hours of curing at standard conditions, the injection material (from the pot life test) is subjected to a concentrated compressive load Ø 20 mm with conical head (top angle 60°) at a rate of 100 mm/min. The load–deformation curve should be recorded.

Reception control should be performed in accordance with the work specification for the rehabilitation work concerned with special control that the mixing time is observed.

Table E2.8   Identification requirements for injection products formulated with hydraulic binder

Basic properties and characteristics	Test method	Tolerance (% deviation from data sheet)
Individual components		
Particle size analysis by laser diffraction[a]	ISO 13320-1	Reference
Identification of freshly mixed product		
Viscosity (Marsh Funnel)[b]	EN 14117	Declared value ±20%
Setting time	EN 196-3	Declared value ±20%
Pot life[c]	EN 9514	Declared value ±20%
Filtration stability[d]	EN ISO 14497	Below prescribed value
Identification of the hardened mixture		
Compression strength and density[e]	EN 12190	Declared value ±15%

[a] The measured particle size distribution should comply with the reference specified by the manufacturer.
[b] Viscosity should be measured 5 minutes after the mixing of the product. The temperature of components before mixing should be 21 ± 2 °C.
[c] The test should be in accordance with EN 9514, but a quantity of 1000 ml instead of 300 ml should be used and three conditioning and test temperatures should be used: 21 ± 2 °C. The minimum use temperature recommended by the manufacturer/supplier with a tolerance of ± 1 °C. The maximum use temperature recommended by the manufacturer/supplier with a tolerance of ± 1 °C. Pot life is defined as the period of time from the mix has been finished until the temperature of the mix has risen 15 °C for polymeric injection materials, and the filtration stability is terminated by the value declared by the manufacturer/supplier for hydraulic injection materials.
[d] Filtration stability is defined as the ability of a hydraulic injection material to pass obstacles (in cracks) and cracks without 'lumping'. This 'lumping' may be caused by: Too large cement particles; poor dispersion; agglomeration (accumulation of cement particles) by injection of fine cracks or by contact with porous areas in the substrate.
[e] The test should be performed after 28 days of curing under standard conditions for hydraulic injection materials.

Table E2.9   Injection products for ductile filling of cracks

Properties and characteristics	Test methods	Compliance criteria
Basic properties		
Effect on polymeric inserts[a]	EN 12637-3	After 70 days, the changes in elongation should be lower than 20% of the initial value
Durability		
Adhesion and elongation after thermal and wet-drying cycles[b]	EN 12618-1 and EN 13687-3	Adhesion. Loss of adhesion lower than 20% of the initial value Elongation: >10%

[a] The contact with polymeric inserts may change the properties of an injection material by a chemical reaction whereby the injection material may lose its ductile properties.
[b] The adhesion and elongation capacity can be reduced by thermal and wet-drying cycling. Under such circumstances, adhesion and failure criteria can be determined as specified in EN 12618-1. Test specimens as specified in EN 12618-1 should be used. They should be submitted 24 thermal and wet-drying cycles according to EN 13687-3.

Table E2.10   Injection products for swelling fitted injection of cracks

Properties and characteristics	Test methods	Compliance criteria
Basic properties		
Effect of polymeric inserts[a]	EN 12637-3	After 70 days, the changes in elongation should be lower than 20% of the initial value
Durability		
Freezing point[b]	ISO 11357-3	Declared value

[a]The contact with polymeric inserts may change the properties of an injection material by a chemical reaction whereby the injection material may lose its swelling properties.
[b]If the freezing point of a swelling injection material is shown by DSC analysis (differential scanning calorimetry), the mechanical properties are determined in function of the temperature by compression testing in the following conditions: cylindrical stamp of diameter, 50 mm; height of the sample, 35 mm; diameter of the sample, 100 mm; rate of compression, 50 mm/min.

Special applications

Where injection materials are used for special purposes, the test methods and compliance criteria shown in Tables E2.9 and E2.10 can be used. Such testing may be required for special projects (with special applications), e.g. where:

- Injection products for ductile filling of cracks come into contact with polymeric inserts or are submitted to thermal and wet-drying cycles. The contact with polymeric inserts may change the properties of the injection material by chemical reaction.
- Injection products for swelling fitted filling of cracks come into contact with polymeric inserts or are submitted to freezing temperatures. The contact with polymeric inserts may change the properties of the injection material by chemical reaction.

References

Mailvaganam *et al. Repair and protection of concrete structures.* Chapter 13: Repair of cracks. CRC Press Inc, FL, USA, 1992.
Turton *et al.* Non-structural cracks in concrete, Report of a Concrete Society Working Party, Concrete Society Technical Report no. 22. Concrete Society, London, UK, 1962.

Annex E3: Changing a crack into a dense joint

Introduction

The standard ENV 1504-9 on 'General principles for the use of products and systems' describes 'Changing a crack into a joint' as method M1.5 and 'Crack covering with local membrane' as method M1.3 in principle P1. By method M1.5 an active (live) crack in a building component is changed into a joint which allows a certain flexibility of the width and keeps the joint watertight, i.e. prevents penetration of aggressive substances. By method M1.3, a crack is covered and tightened by a local membrane. Methods M1.3 and M1.5 deviate significantly from the other methods for protection from penetration of aggressive substances. In ENV 1504-9 the following is stated for these methods:

- These methods use products and systems which are not covered by the EN 1504 standards.

Background for method M1.5

Cracks can be classified into the following groups:

- *Passive* ('dead') cracks, i.e. the cracks are not moving and the cause of cracking is no longer there.
- *Active* ('live') cracks, i.e. the cracks are still moving and the crack width changes and the cause of cracking still exists.
- *Latent* ('dormant') cracks where changes due to, for example, repair revive the crack, i.e. the cause of cracking is present again.

Method M1.5 deals with change of active and latent cracks into joints which can prevent aggressive substances from penetrating into the substrate of the building component. Passive cracks can either be injected by compression or filled with putty before surface protection (Annex E1).

Two parameters dominate the practical application of method M1.5, namely the size of the crack width and the crack propagation. It is a condition for use of method M1.5 that:

- EKP-measurement or measurement of corrosion rate by, for example, Galva Pulse shows that corrosion of the reinforcement of the building component has not begun.
- The concrete is (reasonably) free of chloride ions so that corrosion of the reinforcement cannot be expected to corrode if the situation is unchanged.

- Calculation of moisture shows that moisture will not accumulate in the concrete over the saturation point so that freezing damage or other damage due to moisture in the concrete will not occur.

Method M1.5 may be applied in case of:

- Formation of live cracks on site cast direct-laid concrete floors
- Waterproofing of joints in concrete floor slabs, e.g. in parking houses
- Waterproofing of access balconies composed of elements supported by brackets.

The basis for method M1.5 is that an elasto-plastic bridge is built over an active crack and that this bridge forms a watertight connection between the two sides of the crack. In case of major changes of crack width, the bridge should be very flexible. For minor crack widths, other solutions are possible.

A number of solutions for changing cracks into joints and tightening of joints are on the market. Some of these solutions consist of a combination of steel and plastics, e.g. joints in bridge decks, in parking houses and other heavily trafficked concrete surfaces.

Active crack with small change of crack width

The crack is cut open by a 8–12 mm wide diamond cutter (composed of 2 to 3 blades). The width of the joint should be larger than its depth. This joint does not have to be rectilinear. If the concrete surface is subjected to heavy traffic with large concentrated loads, e.g. industrial floors, it may be necessary to mill off the edges of the joint so that they will not crush due to traffic action.

Then the crack is pointed with a suitable filler and a bridge is built to absorb the movement in the joint. This bridging could be slip tape, so that the deformation of the subsequent membrane can be distributed over an adequate length. Then the membrane is applied, which (while wet) is provided with reinforcement in the form of elastic textile tape. Finally, the concrete surface is given surface treatment (Annex E1).

Active crack with major change of crack width

The crack is cut open as described above. If the concrete surface is subjected to heavy traffic with large concentrated loads, it may be necessary to mill off the edges of the joint so that they will not crush due to traffic action.

If the change of crack width is so large that it is doubtful whether the adhesion of the filler to the substrate can be maintained, slip tape is placed at the bottom of the joint before filling of the joint.

Then the bridging is performed as before, for example, with slip tape so that the deformation of the subsequent membrane can be distributed over an adequate length. Then the membrane is applied, which (while wet) is provided with reinforcement in the form of elastic textile tape. Finally, the concrete surface is given surface treatment (Annex E1).

Tightening of permeable joints

Cast joints between concrete elements are not necessarily tight. Shrinkage and movement of the structure may open these joints so that aggressive substances may penetrate into the substrate. Tightening can be performed by cleaning out the joint, for example, by milling off the edges and treating it as an active crack to be tightened.

This problem is encountered for, for example, surface treatment of access balconies which should be waterproofed to prevent penetration of aggressive substances into the substrate.

Joint with small change of width

In case of a small change of the joint width, a backing can be placed in the joint which is then pointed with a suitably ductile filler, e.g. hardness shore A = 15. Then slip tape is applied over the joint before the membrane is applied, which (while wet) is provided with reinforcement in the form of elastic textile tape. Finally, the concrete surface is given surface treatment (Annex E1).

Joint with major change of width

In case of a major change of the joint width, more flexible bridging should be applied. There are a number of such flexible bridging products on the market, but it is also possible to build a bridge which is very flexible and may contain several barriers to prevent penetration of aggressive substances, see below.

Such solutions may be applied to, for example, waterproofing of access balconies and floors in parking houses.

Background for method M1.3

When the change of crack width is not large, it is often adequate to reinforce with a surface protection over the crack without cutting it open. It may also be sufficient to use a local membrane if only the crack is to be tightened and total surface protection is not necessary. Method M1.3 is a special case of method M1.5.

However, the principles from changing a crack into a joint should be observed. Slip tape should be placed over the crack, and the surface protection (the membrane) should be reinforced with elastic textile mesh (of, for example, glass wool) so that the reinforced membrane can absorb the stress in the membrane due to the growing crack width. The stress is distributed over the width of the slip tape. Knowledge of load–deformation curves for reinforced membrane at different temperatures is therefore necessary.

Manufacturer/supplier: change of a crack into a joint

When membranes or other surface protections are used to change cracks into joints, these materials should fulfil the requirements of EN 1504-2 on 'Surface protection systems for concrete'. Furthermore, there are requirements for other materials beyond the scope of the EN 1504 standards, e.g. shore A for soft fillers.

In particular there is a need for information of the deformability and strength properties of membranes at different temperatures, e.g. freezing temperatures, especially when these membranes are reinforced with elastic textiles. Information of the properties may be in the form of load–deformation curves in the data sheets of the different products. Documentation of load–deformation curves at temperatures of 0, −10 and −20 °C should be available and be included in the data sheets.

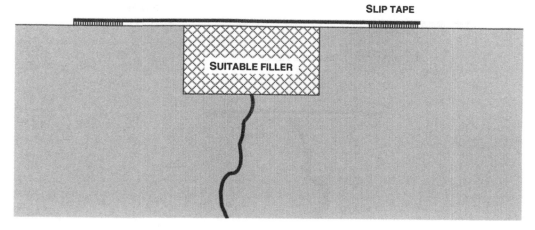

Figure E3.1   Example of change of an active crack with small crack width growth into a joint. Slip tape at the bottom of the joint is not necessary if the filler is sufficiently soft and flexible.

Designer/builder: change of a crack into a joint

The maximum possible crack width is decisive in the design for changing of an active crack into a tight joint. Even though it is possible to calculate the resulting total growth of the crack width, for example, from the prevailing temperature variation, local friction and the crack distribution may lead to uncertainties in the calculations. This should therefore be taken into consideration when determining the design value for the change of joint width.

When the deformations have been determined, the measurement of the necessary width of the slip tape is simple when information of the necessary load–deformation curves of the reinforced membranes at the different temperatures is available (freezing temperatures are often critical for the design).

Contractor/supervisor: change of a crack into a joint

Certain minimum requirements should be fulfilled to comply with the performance requirements made. These requirements are not specified in EN 1504-10 on 'Site application of products and systems and quality control of the works', but they are included in Annex B of EN 1504-10 for information only. Here it is stated that these joints should be designed and performed in accordance with ENV 1992-1-1, other relevant EN standards or standards valid in current use.

Thus there is only little help to be obtained in this way. Therefore, the work specification plays an important role for the execution of the work.

The principles for changing cracks into tight joints are shown in Figures E3.1–E3.3. The filler should be sufficiently soft and have the dimensions 1:2 (Figures E3.1 and E3.2). In case of large movements it is necessary to use slip tape at the bottom of the joint (Figure E3.2). For very large deformations, however, the solutions presented in Figures E3.1 and E3.2 do not apply. A solution for very large deformations is shown in Figure E3.3. Here, a flexible and (mesh) reinforced membrane is applied to the joint against a backing of a split electric pipe. It can, however, be difficult to mount a split pipe, since the width of the crack or joint may vary. Therefore, an ordinary backing would be better. The membrane should also be

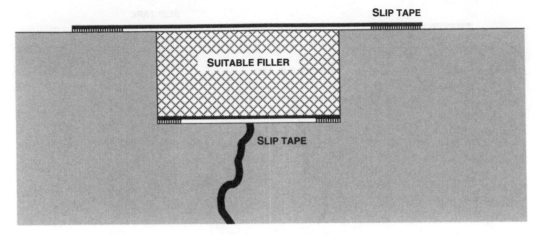

Figure E3.2   Example of change of an active crack with large growth of crack width into a joint. Slip tape is necessary.

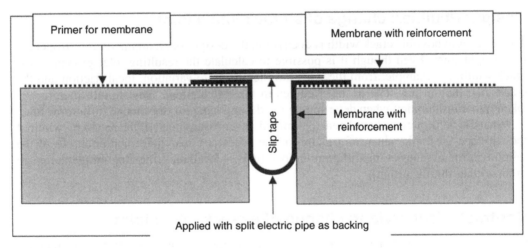

Figure E3.3   Example of change of an active, permeable crack with large growth of crack width into an impermeable joint. It may be difficult to mount a split pipe since the width may vary. It is also difficult to apply a membrane around corners.

applied round the corners, which should therefore be rounded. The interstice is provided with filler and is covered by a membrane on slip tape.

It is general for all three solutions that the applied fillers should not be too soft, since they may be damaged by, for example, stiletto heels and garden furniture.

Annex E4: Structural protection

Introduction

The standard ENV 1504-9 on 'General principles for the use of products and systems' describes principle P1: 'Protection from penetration of aggressive substances' and lists a number of methods related to this principle. The protection should reduce or prevent penetration of aggressive and harmful substances such as water, other liquids, vapour and other gases, radon, chemicals and biological substances. Annex E1 describes methods for surface protection of concrete. However, there are other means of protection, e.g. method M1.6 Structural protection. In ENV 1504-9 this method is described as follows:

- This method applies to products and systems which are not covered by the EN 1504 standards.
- Inclusion of this method in ENV 1504-9 does not automatically entail approval of this method.

Background for M1.6

In many ways structural protection may serve the same purpose as surface protection in the form of hydrophobic impregnation, film depositing paint or membranes. Mounting of panels, large eaves or another form of cover are examples of structural protection. Structural protection may have several functions, for example, to protect from driving rain and insulate (typically gables).

Structural protection may have great influence on the architecture of the building concerned. Today, panels of wood, aluminium, varnished steel plates, etc., are used for structural protection of housing structures. Columns exposed to de-icers at the ground surface can be provided with sleeves of rust-proof steel and reinforced plastics, etc.

Many materials for structural protection are fully impermeable, but for ventilated shielding it is possible to avoid accumulation of moisture and the resulting damage.

Manufacturer/supplier: structural protection

Certain properties of structural protection should be available to assess its applicability as a protection against penetration of aggressive substances. These are not stated in the EN 1504 standards. Since there are a great number of different materials, many standards beyond the EN 1504 series are related to the method M1.6.

Designer/builder: structural protection

Certain properties of structural protection should be available to assess its applicability as a protection against penetration of aggressive substances. Often structural protection should be self-supporting (eaves, panels). Therefore, statistical calculations of structural protection are often dominating. Since there are a great number of different materials, many material standards (e.g. for timber, steel and aluminium) are related to the method M1.6.

Contractor/supervisor: structural protection

Execution of surface protection of concrete surfaces should comply with certain minimum requirements. These requirements are not stated in EN 1504-10 on 'Site application of products and systems and quality control of the works' – but in other material standards.

Annex E5: Electrochemical dehumidification

Introduction

In the standard ENV 1504-9 on 'General principles for the use of products and systems', principle P2 'Moisture control of concrete' is described and number of methods related to this principle are listed. The methods shall reduce the moisture content of the concrete and prevent penetration of new moisture, e.g. water and vapour. Method M2.4 on 'Electrochemical dehumidification' deviates significantly from other methods. In ENV 1504-9 this method is described as follows:

- This method applies to products and systems which are not covered by the EN 1504 standards.
- Inclusion of this method in ENV 1504-9 does not automatically entail approvement of this method.

Background for method M2.4

By electrochemical dehumidification of concrete, an electric difference in potential is applied to the concrete. Thus, alternating voltage is applied to the reinforcement and an electrolytic steel grid is placed on top of the concrete. Also, alternating voltage is applied to the steel grid. The concrete reinforcement should be dominated by positive voltage, which under certain conditions may give rise to reinforcement corrosion. By applying alternating negative and positive voltage to the reinforcement, the tendency towards reinforcement corrosion can be countered, but it requires experience. The risk of rust attack on the concrete reinforcement should be taken into account. The method has not gained footing in Denmark and it has not been possible to obtain any convincing effect by laboratory tests.

Manufacturer/supplier: electrochemical dehumidification

Electrolyte or steel grid or titanium grid as for electrochemical re-alkalization (method M7.5) is used, see EN 14038-1, on 'Electrochemical re-alkalization and chloride extraction treatments for reinforced concrete. Part 1: Re-alkalization'.

Designer/builder: electrochemical dehumidification

There is no experience in Denmark with electrochemical dehumidification and there are no EN standards for the method.

Contractor/supervisor: electrochemical dehumidification

There is no experience in Denmark with electrochemical dehumidification, EN 1504-10 on 'Site application of products and systems and quality control of the works', does not specify requirements for the method and there are no EN standards for the method.

Annex E6: Repair and replacement of damaged concrete

Introduction

The standard EN 1504-3 on 'Structural and non-structural repair' describes repair of damage to concrete building components using mortar or concrete, see principle P3 in ENV 1504-9. The purpose of the repair is to restore the monolithic structure, i.e. strength or aesthetics, of the structure. The principle P3 is here applied fully or in combination with other methods, namely:

- M3.1: Hand filling with repair mortar
- M3.2: Recasting with repair mortar
- M3.3: Application of sprayed mortar or sprayed concrete (shotcreting)
- M4.4: Application of mortar or concrete
- M5.1a: Application of wearing course
- M5.1b: Application of membrane
- M6.1a: Increase of chemical resistance by wearing surface
- M6.1b: Increase of chemical resistance with membrane
- M7.1: Increase of cover thickness with mortar or concrete
- M7.2: Replacement of carbonated or polluted concrete in covers.

There are different requirements for repairing mortar and concrete according to their intended uses. Requirements for repairing mortar and concrete and their application are specified in EN 1504-3 on 'Structural and non-structural repair' and EN 1504-10 on 'Site application of products and systems and quality control of the works', respectively.

Background for the methods M2.1, M3.2 and M3.3

Concrete in building components damaged due to:

- disintegration due to freezing action
- disintegration due to accumulated, penetrating, aggressive substances
- spalling due to rust damage
- spalling due to impacts or abrasion

can normally be removed and replaced by repair. Removal can be performed by cutting off the damaged material and replacing it with repair mortar or concrete. Entire building components can also be removed and replaced by a new element or building component.

The cause of the damage should be removed before or in connection with the repair. Otherwise the possibility of renewed damage is high. In case of frost damage, the repair material should be frost resistant. In case of damage due to rust, the repair material should be able to recreate the passivity of the reinforcement.

ENV 1504-9 on 'Products and systems for protection and repair of concrete structures' divides the principle P3 into four methods according to the method of work:

1   M3.1:   Hand filling with repair mortar
2   M3.2:   Recasting with repair concrete
3   M3.3:   Application of sprayed mortar or sprayed concrete (shotcreting)
4   M3.4:   Replacement of building components, e.g. replacement of a building component or recasting of a new one. This replacement should be designed and performed according to EN 1992-1-1 and EN 206-1 or concrete codes valid in the place of use and not according to ENV 1504-9.

EN 1504-3 on 'Structural and non-structural repair' distinguishes structural repair from non-structural repair. For structural repair the durability and load-carrying capacity (strength) of the repaired building component should be restored. In the case of non-structural repair (or aesthetic repair) such qualities as durability and aesthetics (e.g. geometry and texture) should be restored. Often there is no hard-and-fast boundary between structural and non-structural repairs, since structural repairs also should fulfil certain aesthetic requirements in many cases.

The requirements made in EN 1504-3 on 'Structural and non-structural repair' are summarized in the following tables:

• Table E6.1   Performance characteristics for repair products for all and certain, special intended uses. Properties of repair mortars and concrete for all intentional purposes
• Table E6.2   Identification testing. Tolerances on values declared by manufacturer/supplier
• Table E6.3   Requirements for structural repair products
• Table E6.4   Requirements for non-structural repair products
• Table E6.5   Test methods for properties for special intended uses.

Manufacturer/supplier: materials for repair

Repair mortar and concrete should fulfil certain properties to be applicable for repair of building components of reinforced concrete (Table E6.1). Thus, EN 1504-3 on 'Structural and non-structural repair' specifies requirements for documentation of certain properties of repair mortars and concrete (Table E6.2). Furthermore, minimum requirements are made for these (Tables E6.3 and E6.4).

Requirements for repair materials

EN 1504-3 on 'Structural and non-structural repair' comprises the following groups of repair materials:

• cement-based mortar and concrete
• polymer-modified, cement-based mortar and concrete
• polymer-based mortar and concrete.

Three levels of requirements are made for repair products and repair systems:

1   Requirements for repair materials for all intended uses

2   Requirements for repair materials for certain intended uses
3   Requirements for uses where requirements for the properties and characteristics are not adequate, e.g. extreme environmental action, impact action and vibrations. In such cases supplementary requirements are made by the designer.

Table E6.1   Performance characteristics for repair products for all and certain, special intended uses

Repair principle: Repair method:	P3 M3.1	P3 M3.2	P3 M3.3[b]	P4 M4.4	P5 M5.1[a]	P6 M6.1[a]	P7 M7.1	P7 M7.2
Compression strength	■	■	■	■	■	■	■	■
Chloride ion content[j]	■[c]	■[c]	■[c]	■	■[c]	■[c]	■	■
Adhesive bond (pull-off strength)	■	■	■	■	■	■	■	■
Restrained shrinkage/expansion[d]	■	■	■	■	■	■	■	■
Carbonation resistance[e]	■[c]	■[c]	■[c]	■	■[c]	■[c]	■	■
Elastic modulus[f]	□	□	□	■	■	□	□	□
Thermal compatibility, Part 1[g]	□	□	□	□	□	□	□	□
Thermal compatibility, Part 2[g]	□	□	□	□	□	□	□	□
Thermal compatibility, Part 4[g]	□	□	□	□	□	□	□	□
Skid resistance[h]	□	□	~	□	□	□	□	□
Coefficient of thermal expansion[d, i]	□	□	□	□	□	□	□	□
Capillary absorption and impermeability[g, j]	□	□	□	□	□	□	□	□

■, all intended uses; □, certain intended uses; ~, no requirements.
[a]Methods M5.1 and M6.1 are both based on surface protection according to EN 1504-2.
[b]Some test methods can be modified according to their intended uses.
[c]Not required for plain concrete.
[d]Not required if thermal variation is taken into consideration.
[e]Not required if the repair is given surface protection according to EN 1504-2.
[f]Only for reinforced concrete.
[g]Depends on the exposure class of the repair.
[h]Only required on trafficked areas.
[i]Only for PC (polymer concrete/mortar, i.e. polymer formulated concrete/mortar, e.g. epoxy mortar).
[j]The corrosion resistance of embedded metals depends on the chloride ion content, carbonation resistance and water permeability of the repair material.

Principle 3:   Replacement of damaged concrete
Principle 4:   Strengthening of building components
Principle 5:   Improvement of the physical resistance of concrete
Principle 6:   Improvement of the chemical resistance of concrete
Principle 7:   Restoring of reinforcement passivity.

Method M3.1:   Hand application of repair mortar
Method M3.2:   Recasting of repair concrete
Method M3.3:   Application of sprayed mortar or sprayed concrete
Method M4.4:   Application of mortar or concrete
Method M5.1a: Application of membrane
Method M5.1b: Application of membrane
Method M6.1a: Increase of chemical resistance by wearing course
Method M6.1b: Increase of chemical resistance by membrane
Method M7.1:   Increase of cover thickness by mortar or concrete
Method M7.2:   Replacement of carbonated concrete or polluted concrete in covers.

Table E6.2   Identification testing. Tolerances on values declared by manufacturer/supplier

Property	Test method	Conformity criteria
Granulometry of dry components	EN 12192-1	Reference and tolerance
Infrared analysis[a]	EN 1767	Reference[b]
Unrestrained shrinkage/expansion	EN 12617-4	Reference and tolerance
Compression strength	EN 12190	>80% of manufacturer's declared value
Density	EN 12190	Declared value ±5%
Stiffening time[c]	EN 13294	Reference and tolerance
Workability – Thixotropic mortar[d]	EN 13395-1	Declared value ±15%
Workability – Flow of mortar[d]	EN 13395-2	Declared value ±15%
Workability – Flow of concrete[d]	EN 13395-3	Declared value ±15%
Thermogravimetric analysis[e]	EN 1878	Reference
Epoxy equivalent[e]	EN 1877-1	Declared value ±5%
Amine function[e]	EN 1877-2	Declared value ±6%
Pot life[e]	ISO 9514	Declared value ±15%
Volatile/non-volatile matter in liquid components[e]	EN 1768	Declared value ±10%

[a]For all products containing organic polymers.
[b]Check for signs of change in composition.
[c]As an alternative method, the change in workability with time (e.g. for determination of pot life) may be used (EN 13395, parts 1, 2 and 3).
[d]Check for signs of change in composition.
[e]Only for PC (polymeric concrete/mortar, i.e. polymer formulated concrete/mortar, e.g. epoxy mortar).

Manufacturer/supplier of repair materials should be able to document a number of properties and characteristics (data sheets) of the products they sell. These requirements are listed in Table E6.1.

Designer/builder: design

A rehabilitation project should fulfil certain properties so that they are applicable to repair of building components of reinforced concrete. Thus, according to EN 1504-3 on 'Structural and non-structural repair', documentation of conformity with certain minimum requirements for repair should be available. Furthermore, test methods for special intended uses of these properties should be specified (Table E6.5).

The properties of repair mortars or concrete should be specified as characteristic values (i.e. the 5-percentile) or as a declared value with very little probability of being exceeded.

Requirements for design of structural repairs

Statistic calculation of a structural repair should be made according to EN 1992-1-1 and EN 206-1 or concrete codes valid in current use. Durability requirements should be fulfilled (Table E6.1). A non-structural repair, however, should only be designed based on requirements for durability.

Table E6.3   Requirements for structural repair products

Property	Test method	Compliance criteria	
		Strength class R4[g]	Strength class R3[g]
Compression strength (cube strength)[a]	EN 12190	>45 MPa	>30 MPa
Chloride ion content	EN 1015-17	<0.05%	<0.05%
Capillary absorption	EN 13057	<0.5 kgm^{-2}h$^{-0.5}$	<0.5 kgm^{-2}h$^{-0.5}$
Carbonation resistance[b]	EN 13295	$d_k < $ KB	$d_k < $ KB
Elastic modulus	EN 13412	>20 GPa	>15 GPa
Thermal compatibility, freeze/thaw[c]	EN 13687-1		
Thermal compatibility, thunder shower[c]	EN 13687-2		
Thermal compatibility, dry cycling[c]	EN 13687-4	$f_a > $ 2.0 MPa	$f_a > $ 1.5 MPa
Adhesive bond (pull-off strength)[d]	EN 1542		
Restrained shrinkage/expansion[e]	EN 12617-4		
Skid resistance	EN 13036-4	National requirements	National requirements
Coefficient of thermal expansion[f]	EN 1770	Not always required	Not always required

[a]Requirements are made for the mean value with no single value less than 80% of declared value for the repair material. EN 12190 specifies that at least six specimens should be tested to determine the mean value.
[b]The carbonation resistance of a repair material is determined by comparative testing with control concrete. No requirements for carbonation resistance are made for inside repair or repair where carbonation retarding surface protection is used, i.e. for paint.
[c]Choice of method depends on the exposure conditions. When a product satisfies requirements in EN 13687, part 1, it is deemed to satisfy the requirements in EN 13687, parts 2 and 4. The bond capacity should be verified by reference concrete MC(0.45) after 50 cycles, according to EN 1766. Requirements are made for the mean value with no single value less than 75% of the minimum requirements. Furthermore, maximum crack width after the test should be less than 0.05 mm on average with no observations greater than 0.1 mm.
[d]Adhesive bond should be determined by testing of pull-off strength on reference concrete MC(0.45).
[e]According to EN 12617, part 4, adhesive bond in restrained shrinkage/expansion should be determined by testing of pull-off strength on reference concrete MC(0.45) according to EN 1766. No requirements are made if the compatibility has been verified according to EN 13687, part 1. The requirement is not normally made for repair according to method M3.3 (i.e. application of sprayed mortar or sprayed concrete (shotcreting)).
[f]The coefficient of thermal expansion should be determined according to EN 1770. The requirement is not normally made for repair according to method M3.3 (i.e. application of sprayed mortar or sprayed concrete).
[g]Strength classes R3 and R4: The designing technician should choose the strength class for repair based on the following conditions: Table E6.3 applies to structural repair (see Table E6.4 for non-structural repair); conditions on site; choice of repair method, i.e. method M3.1, M3.2 or M3.3 (ENV 1504-9); choice of repair material. It should be borne in mind that normally the repair should have the same strength properties as the substrate, if possible.

Requirements for design of non-structural repairs

Although a repair is non-structural, requirements are made for compression strength of the repair materials (Table E6.3), but compression strength of mortar and concrete is not the critical property. Properties such as shrinkage, creep, coefficient of elasticity, coefficient of

Table E6.4   Requirements for non-structural repair products

Property	Test method	Compliance criteria	
		Strength class R2	Strength class R1
Compression strength (cube strength)[a]	EN 12190	$>15\,MPa$	$>10\,MPa$
Chloride ion content	EN 1015-17	$<0.05\%$	$<0.05\%$
Capillary absorption	EN 13057	$<0.5\,kg\,m^{-2}\,h^{-0.5}$	$<0.5\,kg\,m^{-2}\,h^{-0.5}$
Carbonation resistance[b]	EN 13295	$d_k < KB$	No requirements
Elastic modulus	EN 13412	No requirements	No requirements
Thermal compatibility, freeze/thaw[c]	EN 13687-1	$f_a > 1.0\,MPa$ after 50 cycles	No cracks and delamination
Thermal compatibility, thunder shower[c]	EN 13687-2	”	”
Thermal compatibility, dry cycling[c]	EN 13687-4	”	”
Adhesive bond (pull-off strength)[d]	EN 1542	$fa > 1.0\,MPa$	$f_a > 1.0\,MPa$
Restrained shrinkage/expansion[e]	EN 12617-4	$fa > 1.0\,MPa$	No requirements
Skid resistance	EN 13036-4	No requirements	No requirements
Coefficient of thermal expansion[f]	EN 1770	No requirements when thermal compatibility is verified	No requirements when thermal compatibility is verified

[a]Requirements are made for the mean value with no single value less than 80% of declared value for the repair material. EN 12190 specifies that at least six specimens should be tested to determine the mean value.
[b]The carbonation resistance of repair material is determined by comparative testing with control concrete. No requirements for carbonation resistance are made for inside repair or repair where carbonation retarding surface protection is used, i.e. for paint.
[c]Choice of method depends on the exposure conditions. When a product satisfies requirements in EN 13687, part 1, it is deemed to satisfy the requirements in EN 13687, parts 2 and 4. The bond capacity should be verified by reference concrete MC(0.45) after 50 cycles, according to EN 1766. Requirements are made for the mean value with no single value less than 75% of the minimum requirements. Furthermore, maximum crack width after the test shall be less than 0.05 mm on average with no observations greater than 0.1 mm.
[d]Adhesive bond should be determined by testing of pull-off strength on reference concrete MC(0.45).
[e]According to EN 12617, part 4, adhesive bond in restrained shrinkage/expansion should be determined by testing of pull-off strength on reference concrete MC(0.45) according to EN 1766. No requirements are made if the compatibility has been verified according to EN 13687, part 1. The requirement is not normally made for repair according to method M3.3 (i.e. application of sprayed mortar or sprayed concrete (shotcreting)).
[f]The coefficient of thermal expansion should be determined according to EN 1770. The requirement is not normally made for repair according to method M3.3 (i.e. application of sprayed mortar or sprayed concrete).
[g]Strength classes R1 and R2: The designing technician should choose the strength class for repair based on the following conditions: Table E6.4 applies to non-structural repair (see Table E6.3 for structural repair); conditions on site; choice of repair method, i.e. method M3.1, M3.2 or M3.3 (ENV 1504-9); choice of repair material. It should be borne in mind that normally the repair should have the same strength properties as the substrate, if possible. For non-structural repair, the strength and elastic modulus should be less than that of the substrate.

thermal expansion tensile strength, pull-off strength and frost resistance may often be far more significant for durability of the repair.

Durability criteria

The compression strength of repair materials is not a measure for the ability of a repair to achieve durability of the structure. The compression strength is significant to structural

Table E6.5 Test methods for properties for special intended uses

Properties and characteristics	Test method	Control concrete
Chloride ion ingress by diffusion[a]	EN 13396	
Creep in concrete compression[b]	EN 13584-2	
Chemical resistance[a]	EN 13529	
Repair of horizontal, downward surfaces	EN 13395-4	MC(0.45)

[a]Not required when efficient surface protection is included in the repair method used.
[b]For PCC (polymer cement concrete, i.e. cement-based concrete/mortar with plastic additives = polymer modified cement mortar) in structural repair, such test according to EN 13584-2 is not normally required if the design compression strength is taken as 60% of the 28-day strength of the repair material.

Table E6.5 contains an overview of supplementary test methods that may apply for special intended uses. Generally, normal repair materials and systems will not be tested according to the test methods mentioned.

However, for special repair projects the above-mentioned test methods may be relevant. Thus, typical examples are situations where documentation of special properties of repair materials and systems is required under the following extreme conditions:

- High or low temperature
- Sea water action
- Action of high salt concentration, e.g. road de-icers with chloride containing de-icing salts.

repairs, but the shrinkage, coefficient of elasticity, coefficient of thermal expansion, creep and tensile strength and adhesive bond strength to the substrate are the significant parameters for failure of the repair and thus reduction of the durability. Use of a shrinkage compensating repair mortar may be an advantage, but a whole group of properties will have to be evaluated prior to choosing such a repair mortar. For protection of chloride penetration, the repair mortar should be estimated based on the chloride penetration parameters.

Choice of repair mortar

The type of repair mortar should be chosen from such properties as strength and durability. Normally, polymer formulated repair mortar should not be chosen without thorough investigation and consideration. Polymer formulated repair mortars have the following properties, among others (compared to cement formulated mortars):

- impermeability to water vapour diffusion
- large coefficient of thermal expansion
- low resistance to fire
- large abrasion strength
- fast strength increment
- can be applied as thin layers
- no finishing treatment (e.g. water storage) is necessary.

Requirements for design of structural repairs

Since a structural repair should be able to transmit mechanical stresses, it should be measured for the purpose, and the only parameters are those of the repair mortar. The design should be in accordance with EN 1992-1-1 and EN 206-1 or concrete codes valid in current use.

First, cracks in the repair mortar or defective bonding on the substrate (i.e. debonding along the edge of the repair) should be prevented. This can be achieved by using a repair mortar with sufficiently large tensile strength and adhesive bond strength (on the substrate). However, it is also significant (apart from choosing a shrinkage reduced repair mortar) to choose a repair mortar where the coefficient of thermal expansion and the coefficient of elasticity are equal to or less than those of the substrate. Further, as previously mentioned, the tensile strength, adhesive bond and contact with the substrate should be good. Finally, the creep of the repair mortar should be small when subjected to compression, but large when subjected to tension.

Sprayed concrete and sprayed mortar (shotcrete), M3.3

When properly performed, shotcreting will result in compact repair and adhesive bond on the substrate of the structure.

Experience from many years shows long durability of properly performed shotcreting. However, it should be noted that normally shotcrete will have a w/c-ratio of 0.40–0.50. This means that the requirements specified in DS 411:1999 for concrete in aggressive exposure class and extra aggressive exposure class cannot be fulfilled (particularly not by the wet method). This should be considered in practice. Generally, the wet method should not be used for repair.

One of the reasons why shotcrete has proved more durable than other concrete with the same w/c-ratio is probably that pure quartz sand is always used for shotcrete (where alkali reaction has never been observed).

Anchorage of repairs

If repair includes the original reinforcement, an area of approximately 30 mm behind the reinforcement should be cut free to ensure proper bonding of the repair (in fresh and cured condition). If such cutting is not used or if the repair does not include the original reinforcement, the repair should be anchored to the substrate by adequate anchors and grids – especially on horizontal, downward concrete surfaces.

Contractor/supervisor: execution and workmanship

The execution of repairs should comply with certain minimum requirements to fulfil the requirements of EN 1504-10 on 'Site application of products and systems and quality control of the works'. This applies to documentation of the repair materials received, workmanship and finishing treatment.

Requirements for execution and workmanship

Repairs with mortar and concrete should be executed so that:

- the specified repair materials are used
- removal of concrete does not impair the substrate
- the specified cleaning and corrosion protection of the reinforcement are observed
- the specified finishing treatment of the repair, e.g. protection from desiccation, is observed

Reception control

EN 1504-3 on 'Structural and non-structural repair' specifies a number of properties for repair materials which should be fulfilled (within certain tolerances) on reception where identification testing is performed (Table E6.2).

If the repair materials and systems concerned are CE marked, identification testing can be replaced by control of the labelling and marking of the materials concerned.

Preparation of the substrate

Cleaning of substrate and its reinforcement prior to repair should be adequate so that repair materials and systems can be applied and the repair will fulfil the requirements for strength and durability.

When the blast cleaning is performed by air, it should be verified that the air is not polluted by oil or other substances which reduce the bond capacity.

The substrate. Weak and damaged and, if necessary, sound concrete should be removed according to relevant principles and methods of ENV 1504-9. Concrete with microcracks and delamination due to removal of the concrete by cutting should be removed by, for example, sand blasting. The cleaned concrete surface should be inspected visually.

Removal of damaged and weak concrete should be made in such a way that pouring and application of repair mortar can be performed with a minimum of encapsulated air. The angle between the cutting surface and the surface of the substrate should thus be between 90° and 135°. If the boundaries of the repair are cut by diamond cutter or similar leaving a smooth section, the surfaces should be roughened by sand blasting.

If the reinforcement corrosion has propagated over more than 30% of the circumference, concrete in an area of at least 20 mm behind the corroding reinforcement should be removed, where the corrosion is due to carbonated concrete, and at least 30 mm where the corrosion is due to chloride infected concrete. In both cases, however, removal of concrete behind the reinforcement should at least correspond to the largest aggregate size of the repair mortar plus 5 mm.

When non-corroding reinforcement is located in carbonated or chloride containing concrete, this state can be maintained provided electrochemical methods are not used or provided that the concrete is in a dry exposure class, i.e. a relative humidity below 70%.

Cleaning of the substrate. On the area to be repaired, the substrate should be free of dust, loose particles, surface pollution (e.g. chlorides) and substances which may reduce the bonding (e.g. water, ice and oil). Unless the cleaning is made directly before the repair, the cleaned repair area should be protected from additional pollution.

Roughening of the substrate. For the methods where roughening of the substrate is required, the surface structure of the cleaned substrate should be suitable for the chosen repair material and system.

Removal of substrate. For the methods where removal of the substrate prior to repair is required, the following requirements should be fulfilled:

- The extent of removal should be as specified in ENV 1504-9
- The quantity of removed concrete should be as little as possible
- Removal of concrete should not impair the performance and safety of the structure. Thus, temporary bracing may be necessary
- The depth of carbonation and profiles for the chloride content and other pollutions of the substrate should be determined and observed
- The extent of the area of substrate to be removed should be in accordance with the required specifications, including:
  - resistance against penetration of liquids and gases
  - existence and intensity of pollution before and after repair

– depth of penetration of the actual pollution
– depth of carbonation
– corrosion activity of the reinforcement
– reinforcement cover
– the need for compaction of the chosen repair mortar and concrete
– the need for adhesive bond on the substrate
– the need for surface protection of reinforcement.

Preparation of reinforcement

Prior to protection of reinforcement and application of repair materials, the requirements for replacement or supplementing of reinforcement should be determined according to the principles and methods stated in ENV 1504-9. The extent of cleaning, surface treatment, improvement of adhesive bond, etc., should be in accordance with the needs established.

Cleaning of reinforcement. Cleaned reinforcement should fulfil the following requirements:

• Rust, mortar, concrete, dust and other loose particles that may reduce the bond between reinforcement and the repair material to be applied should be removed.
• The whole circumference of the reinforcement should be uniformly cleaned. However, this does not apply to electrochemical methods.
• Unless cleaning of reinforcement is made directly before application of repair materials, the cleaned area of substrate and reinforcement should be protected from additional pollution.
• Cleaning of reinforcement should be made without impairing the reinforcement and the substrate or the ambient concrete and environment.
• When reinforcement is polluted by chlorine or other substances leading to corrosion, the entire circumference of the reinforcement should be cleaned by low-pressure flushing so that chlorides and other substances are moved. However, this does not apply to electro-chemical methods.
• When surface protection of reinforcement is necessary according to method M11.1 with 'coating with active pigments', the reinforcement should be cleaned to at least Sa 2, see ISO 8501-1. When surface protection of the reinforcement is necessary according to method M11.2 with 'electrically insulating coating', the reinforcement should be cleaned to at least Sa 21/2, see ISO 8501-1. In other cases of reinforcement protection the degree of cleaning should be specified according to ISO 8501-1 and should be in accordance with the work specification for the project.

Surface protection of reinforcement. Surface protection of reinforcement should be applied to the entire exposed surface. The surface protection agent should not pollute the substrate if it impairs the adhesive bond between substrate and repair.

Removal of reinforcement. Where reinforcement is to be replaced, the following requirements should be fulfilled:

• The substrate must not be damaged.
• The remaining reinforcement must not be damaged.

Supplementary reinforcement. The supplementary reinforcement should fulfil the requirements in EN 10080 on 'Steel for the reinforcement of concrete'.

To counter the risk of corrosion, the supplementary reinforcement should not be in metallic contact with metal of deviating electric potential. However, this does not apply to the different types of steel reinforcement inclusive of steel reinforcement against reinforcement of stainless steel.

When electrochemical methods are applied for rehabilitation, the supplementary reinforcement should be in metallic contact with the existing reinforcement according to the chosen method of rehabilitation.

Improvement of adhesive bond

The improvement of adhesive bond should be specified.

Water for humidifying the substrate should fulfil the purity requirements in EN 206-1 on 'Concrete performance, production, placing and compliance criteria', unless otherwise specified.

Application of repair materials and systems

The repair materials and systems used should be suitable for the actual substrate and able to repair and protect the substrate and its reinforcement, as well as fulfil the requirements in ENV 1504-9.

Repair materials and systems should be stored in such a way that their properties are not impaired. The site conditions should be so that repair materials and systems can be improved, prepared and used in accordance with the standards EN 1504-3, ENV 1504-9 and EN 1504-10.

Before and during application of the repair materials and systems concerned, account should be taken of the temperature and moisture content of the substrate and the characteristics of the ambient air (temperature, relative humidity, dew point and change of humidity due to precipitation and wind).

Preparation and use of repair materials and systems should be in accordance with EN 206-1 on 'Concrete performance, production, placing and compliance criteria', unless otherwise specified. The layer thickness applied should comply with EN 1504-10 unless otherwise specified.

Hand application of repair mortar, M3.1

The substrate should be thoroughly wet (i.e. water saturated and surface dry) at the time where cement-based repair mortars and systems are applied.

Repair mortar should be applied to the prepared substrate and should be prepared so that it can be compacted without encapsulating air. Furthermore, the repair mortar should be able to achieve the required strength and protect the reinforcement from corrosion.

It should be planned how hand application of mortar can be performed without 'sagging' so that the necessary adhesive bond on the substrate is achieved. When interruption of the hand application of repair mortar cannot be avoided so that wet-on-wet cannot be used, a bond improving agent should be used before the work can be continued.

Recasting of concrete, M3.2

The substrate should be thoroughly wet (i.e. water saturated and surface dry) when cement-based repair concrete is applied.

Concrete should be cast in such a way that bleeding of the concrete does not take place. The formwork should fulfil the requirements in standards ENV 1992-1-1 and EN 206-1.

Formwork should be mounted as quickly as possible after preparation and wetting of the substrate. The formwork should be covered and protected from pollution until use.

Concrete to be compacted by vibration should be compacted in such a way that no air is encapsulated. Form cloth may be applied. Further, the concrete should be able to attain the required strength and protect the reinforcement from corrosion.

In the case of repair by self-compacting concrete, the following requirements should be fulfilled:

• The substrate should be prepared as specified above.
• The formwork should be impermeable and also form an impermeable connection to the substrate. Furthermore, the formwork should be shaped so that there is no unnecessary obstacle to the gravitational compaction of the concrete and air should have free access to escape from the concrete.
• The concrete should be cast in such a way that air can escape without vibration.

Sprayed mortar and sprayed concrete (shotcrete), M3.3

Sprayed mortar and sprayed concrete for repair may be performed by the dry method or the wet method. Sprayed mortar and sprayed concrete for repair shall fulfil the requirements in the standard EN 14487-1 on 'Sprayed concrete – Part 1: Definitions, specifications and conformity'.

Sprayed mortar and sprayed concrete should be applied at an angle of 90° to the surface of the substrate, but for reinforcement slant spraying from alternating sides should be used so that the rebound is removed. The distance between spraying nozzle and substrate should be between 0.5 and 1 m.

If the applied layer is thicker than 70 mm, it may be necessary to reinforce the repair with a grid to reduce cracking tendencies. Reinforcement should be rigidly restrained so that vibrations do not impair the embedment. If a double grid is to be mounted, this should be performed in two operations to ensure good encasement.

If sprayed concrete or sprayed mortar is to be applied to undamaged or smooth concrete surfaces, the concrete surface should be blast cleaned until exposure of aggregates. Spike cutting with approximately 15 mm deep holes spaced approximately 100 mm is sometimes recommended, but microcracks, which may damage the adhesive bond, may be created. Sprayed mortar and sprayed concrete should not be applied during unsuitable weather conditions without sufficient cover and protection against the weather.

The need for wetting the substrate before application of sprayed mortar or sprayed concrete should be evaluated, since it will, among other things, depend on the system to be applied. Normally, adhesive bond improvement is not necessary.

Sprayed mortar or sprayed concrete should be applied without encapsulation of air and loose rebound. Furthermore, sprayed mortar and sprayed concrete should be able to attain the required strength and protect the reinforcement from corrosion.

Before shotcreting the substrate should be thoroughly cleaned for shotcreting remains and loose rebounds.

When sprayed mortar and sprayed concrete are to be applied in several layers and this cannot be performed wet-on-wet, the work should be performed as for undamaged surfaces and use of a bond improving agent may be considered.

Supplementary preparation of the finishing layer of sprayed mortar or sprayed concrete should not be performed, unless the finishing layer is a non-structural (i.e. aesthetic) repair applied to an already hardened layer.

Freshly sprayed mortar or concrete should be kept wet for at least 10 days after application and be protected from the weather. Sprayed mortar and sprayed concrete have a large binder content. Therefore, finishing treatment is important to avoid shrinkage cracks, so that a durable repair can be obtained.

Finishing treatment

Finishing treatment is required and should be specified when cement-based repair materials and systems are applied. This is for the sake of cracking due to plastic shrinkage, desiccation shrinkage and heat of hydration. The method of finishing treatment and period of treatment should be specified dependent on the product, layer thicknesses and the environment concerned.

Curing compound should not be applied if there are side effects which may reduce the adhesive bond to subsequent repair materials and systems.

Annex E7: Replacement of building components

Introduction

The standard ENV 1504-9 on 'General principles for the use of products and systems' describes principle P3 'Replacement of damaged concrete' and gives a number of methods related to this principle, including method M3.4 on 'Replacement of building components'.

Background for method M3.4

Repair of building components of reinforced concrete requires a substrate of a certain quality. There are limits specifying the lowest acceptable quality of the substrate where repair instead of replacement of the building components concerned will be more be economical, especially for prefabricated building components.

If repair is very comprehensive in relation to the building component concerned, replacement may be considered.

Replacement of building components often requires bracing during replacement. When building components in monolithic structures are replaced, redistribution of stresses may take place, which should be taken into account (for example, by replacement of part of a bridge deck in a box girder or by comprehensive repair of eccentrically loaded columns).

Manufacturer/supplier: replacement of building components

Availability of special products or systems is not required beforehand. In case of prefabricated concrete elements, they should be manufactured in the normal way.

Designer/builder: replacement of building components

The building components concerned (reinforced concrete) should be designed according to the concrete codes ENV 1992-1-1 and EN 206-1 or concrete codes valid in current use.

Contractor/supervisor: replacement of building components

EN 1504-10 on 'Site application of products and systems and quality control of the works' does not specify requirements for the method. The building components concerned should be mounted or cast in accordance with the concrete codes ENV 1992-1-1 and EN 206-1 or concrete codes valid in current use.

Annex E8: Replacement and supplementing of reinforcement

Introduction

The standard ENV 1504-9 on 'General principles for the use of products and systems' describes the principle P4 'Reinforcement of building components' and gives a number of methods:

- M4.1: Replacement or supplementing of reinforcement
- M4.2: Mounting of reinforcement in bored holes, see EN 1504-6 on 'Anchoring products'. EN 1504-6 only lists requirements for glue for fastening of reinforcement in holes in concrete
- M4.7: Post-tensioning with external cables.

The EN 1504 standards do not specify requirements for supplementary reinforcement. Such requirements can be found beyond the EN 1504 system, e.g. in EN 1992-1-1 and EN 206-1 or concrete codes valid in current use.

Background for the methods M4.1, M4.2 and M4.7

Tensile zones in building components are provided with reinforcement to absorb the tensile stresses in the building component. If this reinforcement is damaged (e.g. by corrosion or collision action) or otherwise defective, it should be improved.

Method M4.1

The traditional form of strengthening of the tensile zone of beams and slabs is to break up and remove the damaged reinforcement and then supplement the tensile reinforcement with a relevant number of reinforcing bars. This reinforcement may be welded to the original reinforcement if the original as well as the supplementary reinforcement is weldable. It is also possible to bond the supplementary reinforcement to the original reinforcement. However, this requires that the necessary anchorage length is present. Another way of fastening supplementary reinforcement is to assemble the reinforcing bars with special devices such as long screw nuts with right-handed and left-handed thread.

Table E8.1   Properties and characteristics to be specified for anchorage products (glue and mortar) according to EN 1504-6

Properties	Rehabilitation method M4.2
Pull-out strength	■
Creep[a]	■
Glass transition temperature	■
Chloride ion content	■

[a] For polymeric anchorage products only.
■, for all intended uses.

Normally there are no problems using rust-proof or acid-proof reinforcement for supplementing corrosive reinforcement if only the concrete can provide the necessary passivating effect (i.e. free from carbonation and chloride ions).

Method M4.2

Strengthening of reinforced concrete structures sometimes requires anchors, bolts or reinforcement, which are embedded in bored holes in the substrate. Typical examples are shear reinforcement in the form of hooks to restrain concrete cast on a slab. Cement paste, cement mortar or synthetic resin may be used for embedment of anchors and reinforcement in the substrate. However, due account should be taken of fire action, e.g. by fire insulation.

EN 1504-6 on 'Anchoring products' specifies requirements for such anchorage products. According to EN 1504-6 an anchorage product is defined as: a hydraulic product or polymer product in the form of a liquid or a paste, for embedding reinforcing bars, bolts or anchors in concrete structures (substrate).

Method M4.7

Strengthening of building components can only be performed by mounting tensile cables on the building component and post-tensioning them. Recently, several types of tensile cables for this purpose have been put on the market, e.g. cables in plastic tubes. Both steel cables and plastic cables, e.g. aramid, are available.

In ENV 1504-9 the following comments on the method are given. This method utilizes products and systems not covered by the EN 1504 standards.

Manufacturer/supplier: replacement and supplementing of reinforcement

The EN 1504 standards do not specify special requirements for products and systems for replacement and supplementing of reinforcement. Other standards specify requirements such as the concrete codes ENV 1992-1-1 and EN 206-1 or concrete codes valid in current use. However, EN 1504-6 specifies requirements for hydraulic and polymeric anchorage products (adhesive mortars and glues). Table E8.1 lists requirements for properties to be documented.

Table E8.2   Control by identification testing (deviations from declared values)

Properties and characteristics	Test method	Tolerance[a]
Granulometry of dry components	EN 1767	Declared value ±6%
Infrared analysis[b, e]	EN 12192-1	Corresponding to reference
Compression strength	EN 12190	Declared value ±15%
Stiffening time	EN 13294	Declared value ±30%
Workability	EN 13395-2	Declared value ±15%
Epoxy equivalent[c]	EN 1877-1	Declared value ±5%
Amine function[c]	EN 1877-1	Declared value ±10%
Pot life[d]	ISO 9514	Declared value ±15%

[a] Data to be supplied by the manufacturer.
[b] For all products containing organic polymers.
[c] For epoxy resin products.
[d] For PCs only.
[e] Check for signs of change in composition.

Designer/builder: replacement and supplementing of reinforcement

The building components and the supplementary reinforcement should be designed according to the concrete codes ENV 1992-1-1 and EN 206-1 or concrete codes valid in current use.

There are several applications of supplementary reinforcement which are fastened by reinforcing bars, bolts and anchors. These applications are not specified in standards such as ENV 1992-1-1 (Table E8.2).

Strengthening according to method M4.4 'Application of mortar or concrete' is a typical example. If the thickness of a beam or slab is to be increased by casting an additional layer of concrete on to it (to strengthen the building component) and this layer of concrete should interact with the building component, the stress resultants (shear and tensile forces) in the construction joint should be absorbed by an adequate volume of reinforcement. This reinforcement should be fastened by anchors, bolts or reinforcing bars glued into bored holes in the substrate.

Another typical example is anchorage of glued flat rolled steel or carbon fibre strips (with multi-orientated carbon fibres) with bolts, which are glued into bored holes in the substrate.

The standard EN 1504-6 on 'Anchoring products' does not deal with other fastenings, such as expansion bolts and self-tapping bolts.

Contractor/supervisor: replacement and supplementing of reinforcement

EN 1504-10 on 'Site application of products and systems and quality control of the works' does not specify requirements for the methods. The supplementary reinforcement of the building components are placed and embedded according to the concrete codes ENV 1992-1-1 and EN 206-1 or concrete codes valid in current use.

Table E8.3 Performance requirements for glue for anchorage products (reinforcing bars, etc.)

Properties or characteristics	Test method	Requirements
Pull-off strength	EN 1881	Displacements ≤0.6 mm at load of 70 kN
Chloride ion content	EN 1015-17	≤0.05% of glue volume[b]
Glass transition temperature	EN 12614	≥45 °C or 20 °C the maximum ambient temperature in service, whichever is the higher
Creep under tensile load[a]	EN 1544	Displacement ≤0.6 mm after continuous loading of 40 kN after 3 months

[a] For Portland cement-based glue only.
[b] According to Danish tradition and experience, this value is too high, since the value corresponds to the threshold value for chloride ions in concrete. According to Danish experience, the value should be taken as 0.01–0.02% of the glue volume.

If glues or mortars are used for fastening of reinforcing bars, bolts or anchors, Table E8.3 of the standard EN 1504-6 lists the requirements for identification testing.

However, if the glues and systems are CE marked, identification testing may be replaced by checking the labels and marking the materials concerned.

Annex E9: Reinforcement with fibre composite materials

Introduction

The standard EN 1504-4 on 'Structural bonding' describes reinforcement of building components reinforced by gluing fibre compote materials on to it (principle P4). However, this principle also includes other methods of strengthening:

- M4.1: Replacement or supplementing of embedded reinforcement (welded joint or threaded joint)
- M4.2: Mounting of reinforcement in bored holes in concrete
- M4.3: Adhesive bonding of flat-rolled steel or fibre composites as external reinforcement
- M4.4: Application of mortar or concrete
- M4.5: Injection of cracks, voids and interstices
- M4.6: Filling of cracks, voids and interstices
- M4.7: Post-tensioning with external cables.

This annex deals with strengthening of building components by adhesive bonding to flat-rolled steel or fibre composites and strengthening of building components of reinforced concrete by mortar or concrete. Requirements for glue used for application of flat-rolled steel and fibre composites and mortar and concrete are made in the standard EN 1504-4 on 'Structural bonding'. Requirements for the execution of adhesive bonding of fibre composites and application of concrete and mortar are specified in the standard EN 1504-10 on 'Site application of products and systems and quality control of the works'.

Background for method M4.3

Fibre composites (e.g. fibre reinforced polymers, FRP) have for many years been used in the aviation and motor industries as structural materials. In the building industry, polymers have mainly been used for protection of structural members against aggressive environments, e.g. membranes in the chemical industry.

However, the need for reinforcement (and upgrading) of building components is growing. Buildings may be used for other than their intended purposes. This may lead to requirements for strengthening of building components. Further, it has been shown that not all concrete has satisfactory durability in the environment to which they are exposed. In cases where protective maintenance has not been made, corrosion of reinforcement may result in requirements for strengthening of the building component concerned.

From 1967 to approximately 1990, strengthening of building parts by adhesive bonding of flat-rolled steel was the dominating principle. Today this principle has (almost) been replaced by strengthening with fibre composites.

Reinforcement with fibre composites has many advantages over reinforcement by bonded flat-rolled steel. Advantages are, among others, the high strength, small density and that fibre composites are easily adjustable on site. Until now fibre composites have mainly been in the form of carbon fibre strips, but now some fibre composites are available as cloth. This means that reinforcement of building components of complicated geometry is possible in a simple way. This fibre cloth is mounted as 'tapestry' in synthetic resin glue without major difficulties, for example, on columns with round cross-section.

Causes of reinforcement

Strengthening building components may be performed based on the following considerations:

- to increase the imposed load on the building
- to increase the quantity of reinforcement in the tensile zone in cases of incorrectly dimensioned reinforcement or misplaced reinforcement
- to procure supplementary reinforcement due to other structural use of the structure, for example, around holes in slabs and for slabs where the supporting walls are to be removed
- to replace or supplement damaged reinforcement due to collision action or corrosion
- to improve structural continuity, for example, over construction joints
- to increase the resistance of the building component to earthquakes, for example, the shear capacity of walls.

Assessment of building components prior to reinforcement

Prior to design and application of reinforcement, the building component concerned should be subject to statistical and technological analysis (of concrete properties). Such assessment will require experience with fibre composites. Therefore, participation of a civil engineer with experience from previous reinforcement applications will be an advantage. The information that may be required by a pilot project is:

- Has the state of the building component concerned deteriorated to such degree that its codified load-carrying capacity has been significantly changed and is no longer adequate?
- Is of the building component concerned
- Has the load of the building component concerned been changed significantly since dimensioning?
- What are the human risks, the risks for the owner and the building structure concerned?
- What costs are involved in the reinforcement, i.e. the direct costs, future costs and the temporary disruption of the performance of the building caused by the reinforcement work?
- What costs are involved in replacement of the building component concerned, the direct costs, future costs and the temporary disruption of the performance of the building caused by the reinforcement work?
- What is the estimated lifetime of the building component concerned in its present state?
- Will inspection (site testing) and maintenance be possible?
- How will reinforcement affect the local infrastructure, local business life, safety and environment?

- Has the matter become political?
- How old is the structure concerned and does it have historical significance, or is it perhaps a preserved building?
- What authorities are to approve the reinforcement project?
- Are there any limitations to the implementation of the project?

Use of fibre composites

Generally, it is possible to strengthen all types of structures by fibre composites. With the different fibre composites available today, building elements with plane (e.g. beams, slabs and plates), as well as curved surfaces (circular columns, arches, cylinder shells and double curvature shells) can be reinforced.

One of the advantages of using fibre composites is that they may be mounted while the structure concerned is still in use. Thus, it is normally not necessary to dismantle installations in buildings or stop the traffic on bridges during reinforcement work. Fibre composites have small density and only require light scaffolding.

In the following, structures are divided into housing structures and bridge structures. The reinforcement principle, however, is the same.

Reinforcement of housing structures

The building components which are typically strengthened by fibre composites in housing structures are beams, slabs, columns and walls. The causes of reinforcement are increased loads, structural changes, faulty dimensioning or execution and reinforcement corrosion (principal reinforcement and shear reinforcement). Further, there are examples of more general causes:

- *Increase of imposed loads.* Heavier use of industrial buildings and changes of a structure from a housing structure or an institutional structure into an industrial building are examples showing that the imposed load may be increased on slabs, columns and walls. In such cases the existing reinforcement in the building components imposed with extra loads is not always adequate to fulfil the required codified safety. Therefore, supplementing reinforcement in the form of bonding with fibre composites can be necessary. The required building components should be reinforced in bending and shear. An increase of the load on the building component by 25–27% is not infrequent. In the tensile zone a carbon fibre strip may be glued on beams and carbon fibre sheets or L-shaped carbon fibre strips may be applied in the shear zone. Circular, centrally loaded columns may be wrapped by several layers of fibre cloth (carbon, glass or aramid). Thus, the concrete in the column will be subjected to biaxial compression which will increase the load-carrying capacity. The method is not so efficient for rectangular columns, but is possible. Eccentrically loaded columns should, apart from fibre cloth, also be provided with carbon fibre strips. The load-carrying capacity of slabs subjected to explosion may be increased significantly if aramid fibre cloth is bonded to the slabs.
- *Structural changes.* The structural changes that may take place are, for example, situations where the building component is not treated properly in the serviceability limit state. This may result in insufficient reinforcement causing too large deformations and unintentional cracking. Examples of balcony slabs with too large deformation because of incorrect treatment in the serviceability limit state are seen. Such building components can be reinforced by application of carbon fibre strips in the tensile zone. Faulty design in the ultimate limit state may also result in problems, for example, the shear capacity of beams, especially

prestressed concrete beams. However, such building components may be strengthened by application of carbon fibre sheets and/or L-shaped carbon fibre strips in the shear zone.

- *Misplaced reinforcement.* There are many possibilities of misplacing reinforcement. For some, application of fibre composites may help, but no particular instruction can be given. Examples are cracks in walls, which can be remedied by application of fibre cloth or short carbon fibre strips.
- *Structural damage.* Collision action on columns and beams are frequent examples of structural damage. Collision action on buildings part of prestressed concrete is particularly sensitive, since an ordinary repair is not sufficient if lines and cables are damaged. Other examples are cutting of reinforcing bars by drilling work. Supplementary reinforcement applied by adhesive bonding carbon fibre strips will repair the damage.
- *Corrosion damage.* By penetration of chlorides and carbonation of the concrete, steel reinforcement loses its passivity and start corroding. By long-term corrosion the reinforcement corrodes to such an extent that the necessary reinforcement and the necessary ductility no longer exist. In such cases the reinforcement should be supplemented after adequate repair. Often, the web reinforcement of beams and columns is impaired so that it is necessary to increase the shear capacity. Other examples of special cases (balconies) are situations where reinforcement (hairpin stirrups and striking plates?) in jointed elements (parapet walls) have corroded (Hansen and Poulsen, 2000).
- *Alkali reaction.* A widespread method for passivating alkali reactions is to prevent moisture absorption (due to driving rain and capillary absorption) in the concrete. Generally, surface protection in the form of dense paint is used. The problem is that the expansion of the concrete does not stop abruptly, but continues and subsides gradually. Thus the paint is subjected to tension so that cracking takes place. This means that the paint is no longer dense and efficient. In cases of alkali reactions in bridge piers with circular cross-section it has been suggested to apply fibre cloth and wrap it round the piers. In this way dense surface protection can be achieved, which at the same time will be able to sustain major tension (Lacasse *et al.*, 2001; Ibrahim *et al.*, 2001). Until now only laboratory tests have been performed, but tests on existing structures with alkali reaction are planned (Wigum, personal communication, 2001).

Reinforcement of bridge structures

The building components, which are typically reinforced by application of fibre composites in bridge structures, are beams, bridge decks and piers. Reasons for reinforcement are increased traffic load, faulty dimensioning or execution and finally, corrosion on reinforcement (principal reinforcement and shear reinforcement). Bridges are outdoor structures which are subjected to moisture (driving rain and capillary absorption). This means that in connection with application of fibre composites an analysis of moisture migration in the structure should also be undertaken, since fibre composites are completely impermeable to water vapour diffusion. There are cases where accumulation of moisture and frost has resulted in damage (bursting). However, there are also cases where carbon fibre sheets are bonded to the soffit of a bridge slab without subsequent damage (Anon, 1997). Furthermore, reinforcement may be required for more general reasons:

- *Increased traffic load.* Heavier traffic (cars and trains) than the bridges were originally designed for. This may apply to bridge slabs, beams and piers. Heavier trains will require reinforcement in the form of, for example, pedestrian tunnels and bridges.
- *Structural changes.* Compared with housing structures, it is rare that a bridge structure is changed. In connection with electrification and transition to double tracks in

southern Jutland in Denmark, reinforcements were necessary. They were achieved by application of flat-rolled steel, since fibre composites were not available in Denmark.

- *Incorrect dimensioning.* It happens that incorrect dimensioning is detected by distinctive cracking as for example on certain slab bridges in the UK in the 1960s. They were not properly designed for torsion due to inclined loading. They were reinforced by application of flat-rolled steel, since fibre composites were then unavailable in the UK.

- *Misplaced reinforcement.* In the UK there are a number of examples that the reinforcement in motorway bridges is not placed correctly. This has called for reinforcement by application of flat-rolled steel and fibre composites. A motorway bridge (Barnes bridge) in Manchester was reinforced by carbon fibre strip because the lap lengths of the original reinforcement were not considered adequate (Sadka, 2000).

- *Corrosion of reinforcement.* By penetration of especially chloride ions (de-icing salts) steel reinforcement loses its passivity and starts corroding. After long-term corrosion the reinforcement corrodes so that the necessary reinforcement area and ductility no longer exists. In such cases the reinforcement should be supplemented after adequate repair. Often, the web reinforcement of beams and columns is impaired so that it is necessary to increase the shear capacity. Columns are particularly exposed due to their location close to the main artery. Normally chloride ions are removed from the concrete of the column, whereupon the column is repaired and reinforced by application of wrapped fibre cloth (of carbon, glass or aramid) (Lynch *et al.*, 1998). The epoxy glue is fully impermeable to chloride ions (and oxygen) so that it forms adequate protection from chloride ions and corrosion in the future. Laboratory tests have been made with corroding reinforcement where the wrapped fibre cloth was applied to the columns without repair (Debaiky *et al.*, 2001). The main conclusion was that total application and wrapping of fibre cloth reduced the corrosion rate significantly, while partial wrapping of the columns did not change the corrosion conditions.

Fibre composites

Building materials have often been based on fibre composites. Clay mixed with stray and sun dried is a typical fibre composite. The best properties of sun dried clay are combined with the properties of the straw resulting in a building material which is better than sun dried clay only. Modern fibre composites consist of high-performance fibres embedded in synthetic resin. Today fibre composites are available as fibre strips and fibre plates, which are applied by adhesive bonding to building components or which are used as backing when concrete is cast, or as fibre cloth, which is bonded to building components (as 'tapestry') in synthetic resin, e.g. for wrapping of columns.

Fibre composites are generally governed by the type, intensity and orientation of the fibres, but also the matrix of the fibre composites are significant for the properties.

Properties of the fibres

In addition to high tensile strength and coefficient of elasticity, fibre composites have different properties which should be carefully considered before choosing fibre composites for a reinforcement project. The following description and properties refer to the fibres. These properties will be somewhat adjusted in case of embedment in synthetic resin for fibre composites.

- *Chemical resistance.* Carbon fibres and aramid fibres are resistant to chemical attacks which may occur in the building industry. However, many types of glass fibre will be attacked by an alkali environment (pH $\geqslant$ 11), but not by acid. Aramid fibres may absorb moisture which is assumed to harm the adhesive bond between fibres and synthetic resin.

- *Resistance to UV-radiation.* Carbon fibres and glass fibres are not sensitive to UV-radiation. Aramid fibres change colour and their tensile strength is reduced when they are exposed to direct sunlight. When the fibres are embedded in synthetic resin, however, only the outer-most fibres will be exposed to UV-radiation. However, synthetic resin will become brittle when exposed to direct sunlight. Therefore, most fibre composites are protected from UV-radiation by a specially protective paint.
- *Electric properties.* Glass fibres and aramid fibres are not conductive. Therefore, they are appropriate for use close to live wires. Carbon fibres are conductive, however, the Japanese standard does not prohibit use of carbon fibre strips near live wires by railway tracks, for example. Carbon fibres should be insulated from the steel reinforcement, but the synthetic resin used for adhesive bonding between carbon fibre strips and reinforcement normally provides adequate insulation. Since carbon fibres are conductive, care should be taken to avoid contact with live wires when the carbon fibre strips are formed on the site.
- *Compressive strength of the fibres.* The compressive strength of carbon fibres and glass fibres is close to their respective tensile strengths. The compressive strength of aramid fibres, however, is considerably lower than their tensile strength.
- *Stiffness of the fibres.* The coefficient of elasticity for carbon fibres corresponds to that of steel, for some types, however, it is significantly higher. The coefficient of elasticity for aramid fibres is lower than that of steel, and the coefficient of elasticity for glass fibres is significantly lower than that of steel.
- *Fire resistance of the fibres.* Glass fibres maintain their tensile strength up to the melting point, which is approximately 1000 °C. In atmospheric air carbon fibres will start oxidizing at approximately 650 °C. Aramid fibres are not applicable at temperatures above 200 °C.
- *Health and safety.* Fibres of carbon, glass and aramid are not injurious to health in normal use. However, protective measures should be taken when composite materials are formed, for example, cut by sawing machine or angle grinder. The fine fibre dust may irritate eyes, mucous membranes and the skin.
- *Environmental aspects.* Fibres of carbon, glass and aramid are not poisonous substances and they are inactive so that they do not liberate harmful substances to the ground water or the atmosphere. Therefore, there is no toxic waste. However, synthetic resin within or for gluing on to fibre composites may present a problem by combustion, since certain synthetic resins may develop toxic fumes.

Fibre strips

When carbon fibres were released for use within the building industry, they were first used for making carbon fibre strip to be applied to building components (of timber, steel, concrete and masonry) as external reinforcement. Adhesive bonding of flat-rolled steel had since 1967 been used for reinforcement of concrete structures. However, there were problems with corrosion. Carbon fibres did not have this problem, but were rather expensive at first. Now, however, the price is more competitive. The first types of carbon fibre strip only had longitudinal carbon fibres. Therefore, bolts could not be used as in, for example, flat-rolled steel, which resulted in problems with anchorage of the fibre strips.

Today carbon fibre strips are available with fibres in more than one direction. They have not yet been commonly used, but are bound to gain wider use due to their better properties.

The properties of this form of carbon fibre strip are close to those of flat-rolled steel, i.e. bolting through the carbon fibre strip down to the substrate is possible.

Carbon fibre strips are manufactured by pulling long 'threads' of carbon fibre through a cleaning bath (i.e. they are cleaned for graphite and are impregnated). Then the carbon fibre threads are pulled through a bath with a suitable number of polymers and through a nozzle with the required dimension. Finally, they are heated so that a matrix for the carbon fibre strip is formed. This matrix ensures that the fibres interact and are able to transfer stresses to the substrate before the gluing process. This process may ensure a carbon fibre intensity of approximately 65%. Carbon fibre strips with both lower and higher intensity may be used depending on the intended purpose.

Carbon fibre strips are manufactured in widths from 40 to 600 mm. The thickness varies from 1.2 to 30 mm and some manufacturers are able to supply carbon fibre strips with a specific thickness according to the customer's requirements. Great lengths of carbon fibre strips are rolled up on coils with a diameter of approximately 1 m.

Fibre plates are manufactured by a similar process, for example, glass and carbon fibre plates for the aviation industry, in widths up to 1.25 m, lengths up to 12 m and thicknesses up to 30 mm. Glass fibre plates typically have fibre intensity of 55% with approximately 10% fibres at an angle of 45° to the main fibres. So far, these fibre plates have not gained widespread use within the building industry.

Fibre sheets

Carbon fibre strips are relatively stiff and are therefore not applicable for building components of complicated geometry. For such purposes carbon fibre sheets are applicable. Fibre cloth is a 'textile' of carbon, glass or aramid fibres. Today there are three types of texture, but more will probably be marketed according to demand:

- Texture with uniaxial fibres
- Texture with fibres in two directions intersecting at right angles. The quantity of fibres may be different in the two directions
- Texture with fibres in three or more directions (multiaxial texture). This is the latest development within fibre cloth. This kind of cloth is especially applicable for shear reinforcement.

Fibre cloth may be used for:

- Reinforcement of building components of complicated geometry, for example, columns with circular cross-section (adhesive bonding and wrapping) (Thériault *et al.*, 2001).
- Absorption of cracking tendency, e.g. due to shrinkage cracks.
- Reinforcement in very thin plates and slabs of high-strength mortar, for example, for corrugated sheets of high-strength mortar (Molter *et al.*, 2001) and for permanent formwork (Brameshuber *et al.*, 2001).
- Preparation of anchorage for adhesive bonding of carbon fibre strips (Sagava *et al.*, 2001).
- Reduction of corrosion rate for columns with corrosive reinforcement (Debaika *et al.*, 2001).
- Reinforcement of columns with alkali reaction and circular cross-section together with reduction of the expansion by total encapsulation (Lacasse *et al.*, 2001; Ibrahim *et al.*, 2001).

Special fibre composites

There are examples of fibre composites of special design:

- Short carbon fibre strips which are preformed into L-shape before hardening. These L-shaped carbon fibre strips have been used for shear reinforcement of beams with rectangular and T-shaped cross-sections (Jensen and Poulsen, 2001), for reinforcement of balconies (Hansen and Poulsen, 2000) and reinforcement of brick chimneys. The strength of L-shaped carbon fibre strips depends on the radius of curvature (Yang *et al.*, 2001). Since the L-shaped carbon fibre strips are assembled by lap joints, such joints are not critical.
- Carbon fibre strips as required by the customer. This applies to thickness as well as distribution of carbon fibres. Carbon fibre strips can be ordered in (almost) any thickness, with parallel fibres in the middle and transverse fibres in the anchorage zones. These are available in Belgium, but not for export (e.g. to Denmark). Such strips are hardly applicable since it goes without saying that they are not stock goods.
- In cases with a great number of repetitive reinforcements it may be an advantage to make special moulds of fibre composites (Anon, 1999).

Synthetic resin for gluing

A general introduction to glue types and their use is given by Mays *et al.* (1992). Epoxy glue is dominant for gluing fibre composites to concrete. There are, however, other types of glue, but they do not have the fine material properties of epoxy glue:

- *Polyester glue*. This type of glue shrinks heavily when it hardens, it has a relatively large coefficient of thermal expansion, is sensitive to alkali environments and has poor bonding to already hardened polyester glue.
- *Vinylester glue*. This type of glue shrinks heavily when it hardens and the pull-off strength (i.e. the tensile strength) is reduced by moisture.
- *Polyurethane glue*. This type of glue shrinks heavily when it hardens and has poor bonding to already hardened polyurethane glue.

Epoxy glues are available in many editions. It will be possible to choose one or more types of epoxy glue that satisfy the requirements for a strengthening project. Some of the most general requirements are:

- *Temperature sensitivity*. Epoxy glue should be able to sustain a maximum temperature of 50 °C without impairment of material properties. The lower glass transition temperature should be $Tg > 50$ °C. Glues that are far less sensitive to temperature than the types up until now have been developed.
- *Fire properties*. When fire safety is significant (e.g. in tunnels), requirements for minimal development of poisonous fumes from the glue during fire should be made.
- *Moisture sensitivity*. When gluing on outside concrete building components it should be checked that the epoxy glue is able to bond to concrete, which is not fully dry (this is significant to strengthening of concrete bridges, especially marine concrete structures).
- *Combined humidity and temperature*. For strengthening of fibre composites in tropical areas and within special industries it should be checked that glue (and matrix) maintains its properties by combined moisture and temperature, for example, above 50 °C and 80% RH.

Manufacturer/supplier: glue for fibre composites

Glue for bonding of fibre composites, as well as primer for application of concrete and mortar, should possess certain properties to be suitable for strengthening of building components of reinforced concrete (Table E9.1). These properties should be specified in data sheets from the manufacturer as characteristic values (5-percentile) or declared values with very little

Table E9.1 Properties and characteristics to be specified according to ENV 1504-9

Properties and characteristics	Principle P4: Structural strengthening[g]	
	Method M4.3[a]	Method M4.4[b]
1. Application		
a. For vertical surfaces and soffits	□	□
b. For upward horizontal surfaces	□	□
c. For injection	□	□
2. Curing		
a. At low or elevated temperature	□	□
b. At humid substrate	~	■
3. Adhesive bond		
a. Composite slab to composite plate	■	~
b. Composite plate to concrete	■	~
c. Corrosion protected steel to corrosion protected steel[d]	□	~
d. Corrosion protected steel to concrete[d]	□	~
e. Hardened concrete to hardened concrete	~	■
f. Fresh concrete to hardened concrete[e]	~	■
4. Durability		
a. Thermal cycling	■	■
b. Moisture cycling	■	■
5. Material properties for designer		
a. Open time[e, f]	■	■
b. Workability[f]	■	■
c. Coefficient of elasticity by compression	■	■
d. Coefficient of elasticity by bending	□	□
e. Compressive strength	~	■
f. Shear strength	■	■
g. Glass transition temperature	■	■
h. Coefficient of thermal expansion	■	■
i. Shrinkage	■	■

[a]Method 4.3 is specified in ENV 1504-9. Normally, satisfactory bonding cannot be achieved when corrosion protected steel is glued to concrete.

[b]Method 4.4 is specified in ENV 1504-9. Adhesive bond of hardened concrete to hardened concrete is typically found between prefabricated concrete elements. Bond of fresh concrete to hardened concrete is typically found in construction joints.

[c]The manufacturer will specify the temperature for the intended use.

[d]In this connection, corrosion protection means application of rust-protecting paint on mild steel.

[e]Does not apply to adhesion by injection technique.

[f]By minimum, standard and maximum use temperatures.

[g]■, properties and characteristics for all intended uses; □, properties and characteristics for certain intended uses; ~, no requirements.

possibility of being exceeded. Further, EN 1504-4 on 'Structural bonding' specifies minimum requirements for these properties (Tables E9.2 and E9.3). In Table E9.1 three types of intended uses are distinguished:

- *General requirements*, to be fulfilled for gluing and casting for strengthening of building components for all intended purposes.

Table E9.2   Properties and characteristics for polymer glues for adhesive bonding

Properties and characteristics	Test method	Requirements[a]
Coefficient of elasticity in bending	EN ISO 178	>2000 MPa
Shear strength	EN 12188	>12 MPa
Open time reference concrete MC(0.40) (EN 1766)	EN 12189	Declared value ±20%
Workability	EN ISO 9514	Declared value[b]
Coefficient of elasticity in compression	EN 13412	>2000 MPa
Glass transition temperature	EN 12614	>45 °C or 20 °C above the maximum use temperature
Coefficient of thermal expansion	EN 1770	<100 × 10^{-6} per °K
Total shrinkage	EN 12617-1	<0.1%
Total shrinkage (alternative test method)	EN 12617-3	<0.1%
Applicable to vertical surfaces and soffits	EN 1799	The glue should not flow more than 1 mm when applied in a thickness less than 3
Applicable to horizontal surfaces	EN 1799	In a flow table test the glue should not spread more than 3000 mm^2 (corresponds to a diameter of 60 mm)
Applicable to injection, MC(0.40) (EN 1766)	EN 12618-2	Failure shall take place in the concrete
Applicable to hardening under special conditions	EN 12188[c]	[d]
Adhesion	EN 12188[c]	Pull-off strength >14 MPa[d]
Durability to fluctuating temperatures and moisture	EN 13733[c]	The shear strength should not be less than the tensile strength of the concrete after thermal cycling or humid-heat exposure[e]

[a]For each batch of the repair material or system according to EN 1504-8, the threshold value corresponds to the mean value for the property concerned.
[b]The workability depends on the quantity and temperature of the sample. It should be noted that the workability of glue is normally less than the pot life of glue.
[c]The test method may require modification for materials other than steel.
[d]The slant shear strength for different values of the angle θ of the slant section should not be less than the tabulated values:

θ (°)	σ_o (MPa)
50	50
60	60
70	70

[e]The steel-to-steel specimen should not fail after exposure to fluctuating humid-heat action.

- *Special requirements*, to be fulfilled for gluing and casting strengthening of building components for special intended purposes.
- *Supplementary requirements*, where the above requirements are not adequate, for example, for strengthening in extreme environments, cryogenic uses, by fire action and impact action (i.e. traffic, ice pressure and earthquakes).

Table E9.3   Properties and characteristics for glue for mortar and concrete

Properties and characteristics	Test method	Requirements[a]
Coefficient of elasticity in bending	EN ISO 178	>2000 MPa
Compression strength	EN 12190	>30 MPa
Shear strength	EN 12188	>6 MPa
Open time reference concrete MC(0.40) (EN 1766)	EN 12189	Declared value ±20%
Workability	EN ISO 9514	Declared value[b]
Coefficient of elasticity in compression	EN 13412	>2000 MPa
Glass transition temperature	EN 12614	>45 °C or 20 °C above the maximum use temperature
Coefficient of thermal expansion	EN 1770	$<100 \times 10^{-6}$ pr. °C
Total shrinkage	EN 12617-1	<0.1%
Total shrinkage (alternative test method)	EN 12617-3	<0.1%
Applicable to vertical surfaces and soffits	EN 1799	The glue should not flow more than 1 mm when applied in a thickness less than 3
Applicable to horizontal surfaces	EN 1799	In a flow table test the glue should not spread more than 3000 mm² (corresponds to a diameter of 60 mm)
Applicable to injection, MC(0.40) (EN 1766)	EN 12618-2	Failure should take place in the concrete
Applicable to hardening under special conditions, MC(0.40) (EN 1766)	EN 12636[c]	[d]
Applicable to hardening under special conditions (alternative test method), MC(0.40) (EN 1766)	EN 12615	The test should lead to concrete failure
Adhesion, MC(0.40) (EN 1766)	EN 12188[c]	For hardened concrete-to-concrete, testing of bond strength should result in concrete failure. For fresh concrete-to-hardened concrete, testing of bond strength should result in concrete failure
Adhesion (alternative test method), C(0.40) or MC(1.40) (EN 1766)	EN 12615	Slant shear strength testing should result in concrete failure
Durability to fluctuating temperatures and moisture, MC(0.40) (EN 1766)	EN 13733[c]	The shear strength should not be less than the bond strength of the concrete after thermal cycling or humid-heat exposure[e]

[a] For each batch of the repair material or system according to EN 1504-8, the threshold value corresponds to the mean value for the property concerned.

[b] The workability depends on the quantity and temperature of the sample. It should be noted that the workability of glue is normally less than the pot life of glue.

[c] The test method may require modification for other materials than steel.

[d] The steel-to-steel specimen should not fail after being exposed to fluctuating temperatures or humid-heat action.

Builder/designer: strengthening of building components

A strengthening project should fulfil certain requirements to result in adequate strengthening of building components of reinforced concrete. Thus, EN 1504-4 on 'Structural bonding' specifies minimum requirements for structural bonding of fibre composites and for concrete and mortar for casting. All of these requirements should be specified and documented.

Properties of glue for adhesive bonding and concrete and mortar for casting should be specified as characteristic values (5-percentile) or declared values with very little possibility of being exceeded.

Requirements for design of strengthening

Statistical calculation of strengthening a building component by adhesive bonding with fibre composites or by application of mortar or concrete should be performed according to the European concrete code ENV 1992-1-1: 'Eurocode 2, design of concrete structures'.

Safety by calculation

So far no codes for strengthening building parts of, for example, reinforced concrete are available. There are, however, a number of instructions:

- *American Concrete Institute*. ACI 440, Guide for the design and strengthening of externally bonded FRP systems for strengthening concrete structures. Draft report by ACI Committee 440, USA, 1999.
- *The Highway Agency*. Strengthening concrete bridge supports using fibre reinforced plastics. Draft Interim Advice Note, UK, 2000.
- *EUROCOMP*. Design code and handbook, Structural design of polymer composites (Clarke, J.L., ed.), E&FN Spon, London, 1996.
- *JSME Concrete Committee*. Recommendation for upgrading of concrete structures with the use of CFRP sheet, Japan, 2000.

Furthermore, instructions for design and use of their products have been worked out by manufacturers of fibre composites.

The partial safety factor method. The partial safety method should form the basis of safety in the ultimate limit state and the serviceability limit state. Assessment of the safety of strengthened building components of reinforced concrete should be made by calculation, testing or a combination thereof. For safety assessment by calculation, requirements are made for:

- Materials, i.e. fibre composites and the adhesive used
- Strength and stiffness of the materials
- Execution of the calculations
- Execution of work
- Control of materials and the works.

Calculations are based on models with realistic reproduction of the performance of the strengthened building component. The technical theory of elasticity is applicable with the generally recognized approximations.

Partial safety factor for (reinforced) concrete. The characteristic compressive strength of concrete is determined by site testing, e.g. a CAPO-test (Poulsen and de Fontenay, 1994) or on bored cylinders (Annex A). Therefore, the partial safety factor for concrete is assumed to be (DS 411-1999 items 5.3.2 (1)P and 3.2.4 (13)P):

$$\gamma_c = \gamma_0 \cdot \gamma_1 \cdot \gamma_2 \cdot \gamma_3 \cdot \gamma_4 \cdot \gamma_5 = \gamma_0 \cdot 1.0 \cdot 1.5 \cdot 1.0 \cdot \gamma_5 = 1.5 \cdot \gamma_0 \cdot \gamma_5 \tag{1}$$

which for normal safety class and normal control class renders $\gamma_c = 1.5$.

Therefore, the design compressive strength is:

$$f_{cd} = \frac{f_{ck}}{\gamma_c}$$

Partial safety factors for (existing) steel reinforcement. In design calculations of ultimate limit states for a strengthened building component, the partial safety factor for steel reinforcement is generally determined by (DS 411, 1999):

$$\gamma_s = 1.30 \cdot \gamma_0 \cdot \gamma_5 \quad \text{for steel reinforcement}$$

where γ_0 and γ_5 take account of safety class and control class, respectively, as follows:

$$\gamma_0 = \begin{cases} 1.10 \text{ for high safety class} \\ 1.00 \text{ for normal safety class} \\ 0.90 \text{ for low safety class} \end{cases}$$

and

$$\gamma_5 = \begin{cases} 1.10 \text{ for reduced control class} \\ 1.00 \text{ for normal control class} \\ 0.95 \text{ for strict control class.} \end{cases}$$

Partial safety factors for fibre composites. The design value of the tensile strength of a fibre composite is taken as:

$$f_{fd} = \frac{f_{fk}}{\gamma_f}$$

where f_{fk} is the characteristic value of the tensile strength of the fibre composite and γ_f is the partial safety factor.

The Eurocode does not specify the value of the partial safety factor for composites. However, EUROCOMP Design Code (1996) gives guidelines for determining the value of the partial safety factor for fibre composites. These guidelines are described in the following.

The partial safety factor is dependent on the type of fibre, the method of production and the time dependence on the coefficient of elasticity of the fibre composite, as well as safety class and control class, so the value may be assumed as:

$$\gamma_f = \gamma_F \cdot \gamma_M \cdot \gamma_E \cdot \gamma_0 \cdot \gamma_5$$

where γ_0 and γ_5 are specified above and where γ_f takes account of the fibre type:

$$\gamma_F = \begin{cases} 1.4 \text{ for carbon fibres} \\ 1.5 \text{ for aramid fibres} \\ 3.5 \text{ for glass fibres.} \end{cases}$$

The coefficient γ_M takes account of the method of manufacture, i.e.:

$$\gamma_M = \begin{cases} 1.1 \text{ for pultrusion and pre-embossing of carbon fibre strips} \\ 1.2 \text{ for embossing of carbon fibre strips} \\ 1.1 \text{ for applying plates by machine} \\ 1.2 \text{ for applying plates by vacuum} \\ 1.4 \text{ for applying plates by hand.} \end{cases}$$

The coefficient γ_E should take into account that the coefficient of elasticity for composites is dependent on time. This may be done by assuming:

$$\gamma_E = \begin{cases} 1.1 \text{ for carbon fibres} \\ 1.2 \text{ for aramid fibres} \\ 1.8 \text{ for glass fibres.} \end{cases}$$

Then the following partial safety factor for tensile strength is obtained for a carbon fibre strip:

$$\gamma_f = \gamma_F \cdot \gamma_M \cdot \gamma_E \cdot \gamma_0 \cdot \gamma_5 = 3.5 \cdot 1.4 \cdot 1.8 \cdot \gamma_0 \cdot \gamma_5 = 8.8 \cdot \gamma_0 \cdot \gamma_5$$

In Denmark it is customary to use a safety partial factor for the tensile strength of carbon fibre strips which is equal to that of steel reinforcement multiplied by the factor φ to take into consideration that carbon fibre strips have brittle failure, i.e.:

$$\gamma_f = \varphi \cdot 1.30 \cdot \gamma_0 \cdot \gamma_5$$

The factor φ is taken as the ratio between the partial safety factors for plain and reinforced concrete, i.e.:

$$\gamma_f = \varphi \cdot 1.30 \cdot \gamma_0 \cdot \gamma_5 = \frac{2.50}{1.62} \cdot 1.3 \cdot \gamma_0 \cdot \gamma_5 = 2.0 \cdot \gamma_0 \cdot \gamma_5$$

which is approximately 15% larger than the partial safety factor according to EUROCOMP (1996).

Anchorage of carbon fibre strips

It appears that one problem in particular should be solved for optimal use of strengthening by adhesive bonding of carbon fibre strips, namely anchorage. During the 5-year period for which carbon fibre strips have been marketed in Denmark, a number of devices have been developed for improving anchorage. Carbon fibre strips may be utilized 100% when these devices are used.

Free anchorage of carbon fibre strips. If a carbon fibre strip is glued on a building component without anchorage improving devices, it is defined as 'free anchorage' on the carbon fibre strip. The ultimate anchorage force is determined by an empirical design method put forward by Deutsche Institut für Bautechnik (1997). However, there are certain restrictions to the German formula (1997): In order to apply the design method, the pull-off strength of the substrate should be a minimum of 1.5 MPa and the maximum pull-off strength to be used in the formula is 3.0 MPa.

Normally, the ultimate anchorage force at free anchorage of a 50 mm wide and 1.2 mm thick carbon fibre strip will be in the range of 25 kN. However, the characteristic tensile strength of such a carbon fibre strip is approximately 165 kN. Therefore, the utilization of carbon fibre strips at free anchorage will be very small. Thus anchorage improving devices have been developed. However, it should be taken into account that most carbon fibre strips are made from carbon fibres, which are parallel with the longitudinal direction of the strip and, therefore, there is no transverse tensile strength of the carbon fibre strip. It is therefore impossible to use bolts through the carbon fibre strip, which was generally used for anchorage with bonded flat-rolled steel.

Recently, carbon fibre strips with transverse carbon fibres have been manufactured (Mathys and Blontrock, 2001), but most carbon fibre strips today have no transverse fibres.

The German formula

Flat-rolled steel and carbon fibre strips bonded to the tensile side of a building component to supplement existing reinforcement are subjected to tensile and shear stresses. In the ultimate limit state it may be assumed (on the safe side) that the bonded reinforcement in building components (beams and slabs) subjected to bending will segregate from the concrete and only be restrained by anchorage forces at the ends, i.e. the model of a reinforced building component is a 'compressive arch and tensional bar'. If shear failure is not possible (EC2) the shear stresses along the external reinforcement should be taken into account so that the anchorage force is reduced.

The ultimate anchorage force F_{fu} of bonded flat-rolled steel or carbon fibre strips with free anchorage depends on the properties of the concrete as well as their own as follows:

$$\frac{F_{fu}}{F_{max}} = \begin{cases} \frac{l_f}{l_{cr}} \cdot \left(2 - \frac{l_f}{l_{cr}}\right), & \text{for } l_f \leq l_{cr} \\ 1, & \text{for } l_f \geq l_{cr} \end{cases} \quad \text{[non-dimensional]}$$

where:

l_f is the applied anchorage length
l_{cr} is a critical anchorage length, see below
F_{max} is the maximum attainable anchorage force at free anchorage, see below

$$l_{cr} = 0.7 \cdot \sqrt{\frac{t_f E_f}{f_a}}, \text{ mm}$$

$$F_{max} = \frac{b_f \cdot k_b \cdot k_T \cdot \sqrt{t_f E_f f_a}}{2000}, \text{ kN}$$

$$k_b = 1.06 \cdot \sqrt{\frac{2 - b_f/b_{ef}}{1 + b_f/400}}$$

b_f is the width of flat-rolled steel or carbon fibre strip (mm)
t_f is the thickness of flat-rolled steel or carbon fibre strip (mm)
b_{ef} is the average space between flat-rolled steel or carbon fibre strip, e.g. in a slab (mm)
E_f is the coefficient of elasticity of flat-rolled steel or carbon fibre strip (MPa)
f_a is the site pull-off strength of the cleaned concrete surface (MPa)
k_T is the coefficient of thermal expansion, which is $k_T = 1$ for inside building components and 0.9 for outside building components.

It should be noted that not all of the above formulae are dimensionally correct.

Anchorage improving devices

The tensile strength of carbon fibre strips, in particular, is so large that they cannot be fully utilized by free anchorage. The different manufacturers and suppliers have therefore developed different devices and documented their efficiency:

- *Anchorage by adhesive bonding of transverse strips.* A simple way of increasing the ultimate anchorage force is to apply two transverse strips to the anchorage length of the carbon strip at right angles to the carbon fibre strip. Thus the ultimate anchorage force may be increased to approximately 170% compared with free anchorage (Jensen *et al.*, 1999). A special form of anchorage by transverse strips is to apply one or more layers of carbon fibre sheets.
- *Anchorage with bolt-glued anchorage plates.* A carbon fibre strip may be anchored by one or more bonded anchor plates which are anchored to the substrate. It is important that the bolts are not drilled through the carbon fibre strips unless the carbon fibre strips used have transverse carbon fibres. It is possible to double the ultimate anchorage force compared to free anchorage.
- *Sandwich anchorage.* Carbon fibre strips may be glued between two parallel steel plates, which are thoroughly bolted together and anchored to the substrate. Thus, shear failure in the concrete is avoided. It is replaced by slip failure in synthetic resin glue between steel and carbon fibre strip. Thus the ultimate anchorage force can be rather significantly increased.
- *Wedge anchorage.* A wedge anchorage may be developed based on sandwich anchorage by having the plates form a suitable angle. The plates should be secured by heavy bolts so that a backstop to the tensile stresses in the glue is achieved. Thus, there will be a wedge effect in the joint. This will result in increased ultimate anchorage force, or in other words, that a given anchorage force can be absorbed on a shorter wedge anchorage than by sandwich anchorage.
- *Core anchorage.* A hole of diameter 120–150 mm is bored in the concrete slab where the carbon fibre strip is to be anchored. At the hole the grooved wedge should be at least 50 mm deep and the length at least 300 mm. The carbon fibre strip is provided with a 'handle' of synthetic resin glue and applied to the concrete at the bottom of the bored hole and fixed in this position. The hole is filled with synthetic resin glue or a reinforced high-strength repair mortar, which is steel reinforced (by 2 to 3% by volume) to achieve adequate tensile strength.
- *Bolt-glued anchorage.* Flat-rolled steel and carbon fibre strips with multidirectional carbon fibres may be anchored directly by bolting through the flat-rolled steel or the carbon fibre strip into the substrate. The anchorage force is not transferred directly to the bolt through compression on the bearing and the bolt should be dimensioned for shear (transverse load) (Højlund Rasmussen, 1962).
- *Post-tensioned carbon fibre strips.* Post-tensioning of carbon fibre strips has been developed and was introduced in Denmark in 2002 (Andrä *et al.*, 2002).

Contractor/supervisor: strengthening of building components

Strengthening of building components of reinforced concrete should be based on design material comprising work specification and drawings, etc., complying with the requirements EN 1504-10 on 'Site application of products and systems and quality control'.

It is a condition for performing the work that registration and evaluation of the state of the concrete and reinforcement of the building component are available together with the designer's assessment confirming that it is possible to strengthen the building component.

Thus, documentation of the received repair materials, execution of the work and finishing treatment should be available.

Requirements for execution of work

Strengthening by applying flat-rolled steel and fibre composites by adhesive bonding or by applying mortar and concrete should be performed so that:

- The specified repair materials are used
- Cleaning and levelling of concrete do not harm the substrate
- The specified cleaning and gluing process and application of concrete or mortar are complied with
- Fire protection, when required, is made according to specifications.

Concrete

It is assumed that the concrete is investigated for cracks and defects and that an estimate is available specifying how these defects can be remedied before strengthening is made. Prior to strengthening the defects, cracks and delamination of a building component should be repaired according to ENV 1504-9 on 'General principles for use of repair materials and systems'.

The surface of the substrate should be cleaned and weak concrete should be removed without introducing microcracks or other defects which will reduce the pull-off strength of the concrete surface. However, as little as possible of the concrete should be removed. After repair, the concrete surface should be roughened, e.g. by sand blasting to fulfil the requirements made in the work specification. Finally, the concrete is vacuum cleaned to remove dust, loose concrete and material which may reduce the pull-off strength of the concrete surface.

A serious defect in concrete is delamination. If inspection of the building component has revealed delamination of the concrete, such delamination should be repaired in accordance with ENV 1504-9 on 'General principles for use of repair materials and systems'.

Testing

In connection with execution of the work, the substrate and its reinforcement should be tested as specified in the work specification. This test may, however, be performed earlier in connection with the registration and evaluation of the state of the concrete structure.

Pull-off strength. The cleaned concrete surface of the building component (the adhesive surface) should be evaluated by the pull-off strength which is determined according to test method EN 1542 on 'Pull-off test'. The pull-off strength of a cleaned concrete surface to be bonded to flat-rolled steel or fibre composites should fulfil the requirements made in the work specification. Normally, this will indicate pull-off strength of minimum 1.5 MPa.

Documentation of conformity with the requirements of the work specification for the pull-off strength of the substrate may be performed as follows. The pull-off strength of the cleaned concrete surface is determined at an adequate number of points and the characteristic value is determined (Annex A). It is shown statistically (i.e. by hypothetical testing) that the characteristic value of the pull-off strength of the cleaned substrate is larger than required in the work specification.

Compressive strength. The compressive strength of the concrete (in the compression zone) will normally be determined in connection with inspection and preliminary testing of the building component. Therefore, determination of the compressive strength of concrete according to EN ISO 8046 on 'Determination of pull-out force' (Poulsen and de Fontenay, 1994), is not necessary unless required in the project work specification.

Corrosion state of the reinforcement. Normally the corrosion state of the reinforcement will be determined in connection with the registration of state. Therefore, determination of the corrosion state of the reinforcement is not necessary unless required in the work specification.

If corrosion has been detected in the reinforcement of the building component, the cause should be found. Then the reinforcement should be passivated in accordance with ENV 1504-9 on 'General principles for use of repair materials and systems'.

Reception control

EN 1504-4 on 'Structural bonding' specifies a number of properties of repair materials for strengthening which should be observed within certain tolerances on reception, where identification testing is made (Table E9.4).

For CE-marked repair materials and systems, a check of labels and marking is sufficient.

Mixing of glue

All tools for mixing of glue should be cleaned and in a good state. The personnel should be trained for mixing and application of glue and have attended a course on 'Application of isocyanate epoxy products'.

Normally glue is supplied as two components to be mixed. This should be done and the manufacturer's instruction should be carefully observed. The two components should be carefully mixed. After the first mixture the mix is shifted to a new bucket and the mixing process is continued. Thus, a uniform glue mixture can be achieved.

Application of glue to carbon fibre strips

A 1–2 mm thick layer of glue is applied to the cleaned glue area. Before application of the glue to carbon fibre strips the strips should be cleaned so that they do not rub off on a white cloth. Certain carbon fibre strips are provided with protection strips by the manufacturer. This strip should be removed before gluing.

Then the glue is applied to the carbon fibre strip in a roof-shaped coating so that the glue is pressed to the sides by the application. The thicker coating at the centre line of the carbon fibre strip will ensure that no interstices are formed in the glue.

Application of glue to fibre cloth

For application to fibre cloth (carbon, glass or amide), more high-flowing glue than for flat-rolled steel and carbon fibre strips is used. The glue is applied to the cleaned substrate by brush or roll. The fibre cloth is pressed by the roll so that the glue is visible through the cloth. Then more glue is applied and a new layer of fibre cloth is applied. Normally, up to three layers of fibre cloth can be applied wet-on-wet.

There are special machines for embossing large widths of fibre cloth for gluing.

Lap joints in fibre cloth should be in accordance with the manufacturer's specifications and should normally not be less than 200 mm. The number of fibre cloth layers should not exceed five (FIB task group 9.3), but there are examples of application of 10 layers without complications.

Table E9.4 Requirements for identification testing (identification control)

Properties and characteristics	Test method	Requirements for tolerance
Colour	Visual	Uniform and corresponding to the manufacturer's specification
Granulometry for filler in polymer glues	EN 12192-2	Declared value ±5%
Content of ashes by direct calcination	EN ISO 3451-1	The larger of the values: Declared value ±5% or 1% point of the total product
Thermogravimetry of polymers: Temperature scanner method	EN ISO 11358	The larger of the values: Declared value ±5% or 1% point of the total product
Infrared analysis of glue and hardener	EN 1767	Position and relative intensity for main absorption strips should correspond to the manufacturer's specification
Pot life	EN ISO 9514	Declared value ±20%

The manufacturer should perform representative sampling and identification testing in accordance with this table. This testing should at any time document the composition of the repair material or system concerned and should be repeated under the following circumstances:

(a) When a new composition or a new type of repair material or system is introduced.
(b) When the composition of a repair material or system is changed and this affects the properties and characteristics.
(c) When raw materials used for repair materials and systems are changed and this affects the properties and characteristics.

Information of the acceptable tolerances specified by the manufacturer is given in the table. Control values and other documentation should be kept by the manufacturer.

Performance requirements for structural glues are made in Table E9.2 on 'Properties and characteristics for polymer glues for adhesive bonding' and in Table E9.3 on 'Properties and characteristics for glue for mortar and concrete'. The manufacturer should perform representative performance tests, and the glues concerned should fulfil the requirements in Tables E9.2 and E9.3. The performance test should be repeated under the following circumstances:

(d) When a new composition or a new type of repair material or system is introduced.
(e) When the composition of a repair material or system is changed and this affects the properties and characteristics.
(f) When raw materials used for repair materials and systems are changed and this affects the properties and characteristics.
(g) At maximum 5-year intervals.

Control of the works

In the work specification, the necessary control of the works is specified. If not otherwise specified, the control can be performed in the following way:

● *Strength of glue.* It should be documented that the applied glue is able to attain the required strength. If not required otherwise, this can be done as follows. The bond strength of mixed glue is tested by pull-off testing according to EN 1542 on 'Bond strength by pull-off'. This is done by pulling-off steel roundels glued to a steel plate, which is cleaned to St 2 by a steel brush (e.g. a 20 mm thick steel plate or an HE steel beam no. 2000 Eurocode 53-62). After hardening for 3 days at a temperature of maximum 20 °C, the pull-off strength is tested. A minimum of two pull-off tests should be made for a maximum 10 mixes of the glue. No single value should be less than required in the work specification. In cases where the pull-off strength is less than required in the work specification for the project, the cause should be found (e.g. mixing procedure, exceeding of pot life).

- *Filling of glued joints.* It should be documented that glued joints between applied flat-rolled steel or composites and substrate have no defects such as encapsulated air (voids). This can be done by the test method 'impact echo' or similar testing. 100% control on the outermost third of each flat-rolled steel or carbon fibre strip, unless the anchorage applied is independent of this bond.

Safety and health

ENV 1504-9 on 'General principles for use of repair materials and systems' or EN 1504-4 on 'Structural bonding' do not specify requirements for safety and health apart from the national rules, which should be observed.

Therefore, the repair materials used should fulfil the requirements made in the work specification and the Working Environment Service. This implies, among other things, that repair materials should be available on site in their lawful, labelled original packing.

Special intended uses

Table E9.3 comprises all intended uses and certain intended uses. Moreover, there may be special intended uses, namely for dynamic action (ENV 1992-1-1). Table E9.5 specifies test methods which may be applied to special intended uses, e.g. dynamic action. Normally, two cases of dynamic action are of interest:

- There is dynamic action during the hardening process of the glue. This may be the case for strengthening of a railway bridge where the train traffic cannot be closed down during the hardening process of the glue (normally lasting approximately 3 days).
- There is dynamic action after the hardening process of the glue. This may be the case for strengthening of a railway bridge. Building structures may also be subjected to dynamic action, e.g. in printing houses.

Fire exposure

When gluing external reinforcement on to building components for which requirements for fire resistance are made, the manufacturer should declare the fire class for the hardened glue.

Hardened glue containing more than 1% of homogeneously distributed organic material (by mass or volume, whichever is stricter), should be classified according to EN 13501-1 and the manufacturer should declare the fire class.

Table E9.5   Test method for special intended purposes

Properties	Test methods
Fatigue failure due to dynamic action during hardening	EN 13894-1
Fatigue failure due to dynamic action after hardening	EN 13894-2

Liberation of poisonous gases

Hardened polymer glues should not liberate gases which may be harmful to health, safety and environment. Further information can be obtained from the following website, http://europa.eu.int

References

ACI Committee 440. Guidelines for the selection, design, and installation of fiber reinforced polymer (FRP) systems for externally strengthening concrete structures. USA, 1999.

Adrä, König, Maier. First applications of CFRP tendons in Germany. IABSE Symposium, Melbourne, Australia, 2002.

Anon. Reinforcement of structures with carbon fibres. *Freyssinet Magazine*, 1996/1997.

Anon. Repairing columns without using forms. *Concrete International*, 21(3), 1999.

Debaiky, Green, Hope. Corrosion of FRP-wrapped RC cylinders – long term study under severe environmental exposure. Proceedings of the Fifth International Conference on Fibre-Reinforced Plastics for Reinforced Concrete Structures. Cambridge, UK, 2001.

Deutsche Institutt für Bautechnik. Richtlinien für das Verstärken von Beton-bauteilen durch Ankleben von unidirektionalen kohlendtoffaserverstärkten Kunst-stofflamellen (CFK-lamellen), Typ Sika CarboDur. Zulassungsnummer Z-36.12-29, Berlin, 1997.

EUROCOMP. *Design code and handbook, structural design of polymer composites.* Clarke, J.L. (editor), E&FN Spon, London, 1996.

Hansen, Poulsen. Altanrenovering, Lufthavnsparken Melstedhusene. Arkitekt- og bygge-bladet ark.byg.1:2000, januar.

Jensen, Petersen, Poulsen, Ottosen, Thorsen. On the anchorage to concrete of SikaCarboDur CFRP strips, free anchorage, anchorage devices and test results. International congress, creating with concrete, Dundee, 1999.

Jensen, Poulsen. L-shaped CFRP strips against shear failure. Proceedings of the International Conference on Composites in Construction, Porto, 2001.

JSCE Concrete Committee. Recommendation for upgrading of concrete structures with the use of CFRP sheets. Japan, 2000.

Mathys, Blontrock. Luc Taerwe. Brandgedrag en duurzaamheid van gelijmde wapeningen, uwkroniek Techniek, Maj, 2001.

Molter, Littwin, Hegger. Cracking and failure modes of textile reinforced concrete. Proceedings of the Fifth International Conference on Fibre-Reinforced Plastics for Reinforced Concrete Structures. Cambridge, UK, 2001.

Poulsen, de Fontenay, 1994. Dokumentation og undersøgelse af beton i bygværker, ved 5-års eftersyn, i almindelige syns. og skønssager, før reparation. Bilag 1 om Prøvningsmetoder. Beton 7, Statens Byggeforskningsinstitut, Hørsholm, 1994.

Sadka. Strengthening bridges with fibre-reinforced polymers. *Concrete*, 34(2), 2000.

Sagawa, Matsushitra, Tsuruta. Anchoring method of carbon fiber sheet for strengthening of reinforced concrete beams. Proceedings of the Fifth International Conference on Fibre-Reinforced Plastics for Reinforced Concrete Structures. Cambridge, UK, 2001.

The Highway Agency. Strengthening concrete bridge supports using fibre reinforced plastics. Draft interim advice note, UK, 2000.

Thériault, Claude, Neal. Effect of size and slenderness ratio on the behaviour of FRP-wrapped columns. Proceedings of the Fifth International Conference on Fibre-Reinforced Plastics for Reinforced Concrete Structures. Cambridge, UK, 2001.

Yang, Nanni, Chen. Effect of corner radius on the performance of externally bonded FRP reinforcement. Proceedings of the Fifth International Conference on Fibre-Reinforced Plastics for Reinforced Concrete Structures. Cambridge, UK, 2001.

Annex E10: Increase of cover

Introduction

The standard ENV 1504-9 on 'General principles for the use of products and systems' describes principle P7 'Restoring reinforcement passivity' and gives a number of methods for application, including method M7.1 on 'Increase of cover with mortar or concrete'.

Background for method M7.1

If the concrete of the reinforcement cover is not polluted with chloride ions or other aggressive substances, but 'only' too thin, the cover thickness may be increased by application of an adequately dense (i.e. carbonation retarding or chloride ion retarding) mortar or grout. However, it is a condition that the adhesive bond between mortar/grout and the substrate is ensured for the stipulated lifetime.

In case of lack of cover (e.g. due to collision), the cover should be restored by application of concrete (Annex E6).

Manufacturer/supplier: increase of cover

The EN 1504 standards do not specify special products and systems for increase of cover. The cover may be established by application of elastic grout. The manufacturer/supplier should in this case specify and document the resistance of the elastic group to penetration of chloride ions and carbonation versus the coat thickness.

Typically, the equivalent thickness of elastic grout in relation to carbonation may be 10 mm per mm of coat thickness of the grout at a w/c-ratio of $w/c = 0.6$ for the reference concrete.

Designer/builder: increase of cover

The necessary cover thickness should be designed in accordance with the concrete codes ENV 1992-1-1 and EN 206-1 or concrete codes valid in current use.

Contractor/supervisor: increase of cover

EN 1504-10 on 'Site application of products and systems and quality control of the works' does not specify requirements for the method. The supplementary cover of the building components concerned should be applied according to the manufacturer/supplier's specification.

Annex E11: Re-alkalization by natural diffusion

Introduction

The standard ENV 1504-9 on 'General principles for the use of products and systems' describes principle P7 'Restoring reinforcement passivity' and gives a number of methods for application, including the method M7.4: 'Re-alkalization of carbonated concrete by natural diffusion'. However, there are no special standards in the EN 1504 series or other EN standard describing this method. The method is applied in Germany.

Background for method M7.4

Concrete is a porous and permeable material allowing moisture migration, i.e. the pore water of the concrete. When the carbonation front has almost reached the outermost reinforcement layers in a building component, a cement mortar with very high pH value (14–15) is applied. The mortar is protected from desiccation for an adequate time so that hydroxyl ions may diffuse into the carbonation concrete, causing re-alkalization of the concrete, and the reinforcement in the carbonation zone regains its passivity.

It is known that repair with high-alkaline repair mortar may re-alkalize the adjacent concrete by natural diffusion of hydroxyl ions.

The applied mortar may change the architecture of the building significantly and, therefore, the method will be suitable for already plastered structures.

Manufacturer/supplier: re-alkalization by natural diffusion

Method M7.4 does not require any special products besides dry mortar which has a very high pH value when fresh. This mortar should fulfil the requirements in EN 1504-3 on 'Structural and non-structural repair'.

Designer/builder: re-alkalization by natural diffusion

Protection of the applied mortar should be planned and designed so that the mortar may be kept humid for an adequate period. Proportioning of the mortar should take into consideration that the mortar is open (so that diffusion may occur) and that it does not carbonate in the stipulated lifetime of the building structure.

Contractor/supervisor: re-alkalization by natural diffusion

Execution of surface protection should fulfil certain minimum requirements. These requirements are specified in EN 1504-10 on 'Site application of products and systems and quality control of the works', but need not be very detailed.

Annex E12: Electrochemical re-alkalization

Introduction

The standard EN 14038-1 on 'Electrochemical re-alkalization and chloride extraction for reinforced concrete – Part 1: Re-alkalization' describes method M7.3 i ENV 1504-9 on 'General principles for application of products and systems' dealing with re-alkalization of carbonated concrete in existing, normal concrete structures. The method is applicable to re-alkalization of concrete, with normal, embedded reinforcement which has been subjected to the atmosphere (and therefore carbonated). However, method M7.3 is not applicable to concrete with prestressed reinforcement for high-strength steel which is given epoxy treatment or which is galvanized.

Background for method M7.3

Concrete in building components with penetration of aggressive substances often shows damage. A frequent cause is penetration of carbon dioxide CO_2 so that the concrete carbonates and no longer has rust protecting properties in humid environments.

Carbonation of concrete

Porous concrete carbonates rather quickly due to the carbon dioxide CO_2 of the air (Poulsen *et al.*, 1985). When the cover of the reinforcement is fully carbonated (i.e. is calcified) the concrete loses its rust-protecting effect. Therefore, if the concrete is also exposed to cycling humidity, i.e. in moderate exposure class, the condition for corrosion of the reinforcement will be present. Reinforcement in carbonated concrete will therefore corrode, but only if there is adequate admission of moisture and oxygen to the concrete. Fronts and balconies of too porous concrete are typical examples of structural members likely to corrode due to carbonation of the concrete. However, reinforcement in concrete in passive exposure class (e.g. inside, heated concrete) is not exposed to corrosion due to lack of moisture.

If the w/c-ratio of concrete is appropriately low as prescribed by the concrete code, there will hardly be any problems for a long period (50–100 years). The previous concrete codes, however, prescribed w/c-ratios that today are known to be so high that the lifetime of structures is often less than generally accepted.

Often concrete in, for example, pre-cast construction with narrow dimensions is subjected to reinforcement corrosion due to carbonation, especially if the reinforcement is not placed with sufficient accuracy (variation of cover).

Re-alkalization before corrosion

Carbonation of concrete has caused a considerable amount of concrete damage all over the world. Especially concrete in pre-cast construction from the 1960s and the 1970s has corrosion damage today due to carbonation of the concrete.

By periodical and careful inspection of concrete structures in moderate exposure class (e.g. fronts), however, it is possible to detect initial carbonation before the damage propagates too much. If such initial carbonation is not detected, rehabilitation may be costly, because replacement of reinforcement and concrete or entire building components may result.

Reinforcement corrosion due to carbonation can be extensive without timely intervention. Electrochemical re-alkalization will be costly, just as any other rehabilitation, when the reinforcement has extensive corrosion damage, e.g. resulting in cracking and spalling of concrete.

However, use of electrochemical re-alkalization can often be optimal as part of preventive maintenance.

Electrochemical re-alkalization

Electrochemical re-alkalization of a carbonated concrete surface is theoretically very simple. By electrochemical re-alkalization there are two processes that increase the pH value of the concrete and thus the rust-protecting effect. These two processes are electrolysis at the reinforcement and penetration of an alkaline electrolyte into the concrete surface by electro-osmosis.

Concrete is normally porous and moist. The pH value of the pore water in carbonated concrete should be increased if the reinforcement is re-passivated, i.e. rust protected.

An electrochemical re-alkalization is performed by imposing negative voltage (cathode) on the reinforcement and applying a coat of papier mâché, textile or similar which is moistened by an alkaline electrolyte. This electrolyte may be a solution of calcium hydroxide or soda (sodium carbonate). As anode, a titanium grid (or steel grid) is applied in this electrolyte and a positive voltage is imposed on the grid.

A current of approximately 1 A per m^2 of concrete surface is supplied and over a period of 4 to 7 days (dependent on the conditions) the concrete will re-alkalize by electrolysis and electro-osmosis.

Electrolysis

In case of reinforcement on which negative voltage is imposed, hydroxide ions are produced by electrolysis, the so-called cathodic reaction.

Due to the hydroxide ions the pH value of the concrete rises to above 14 around the reinforcing bars. Thus the reinforcement recovers its passivity, i.e. it is now protected from corrosion due to carbonation. Roughly, these alkaline zones propagate concentrically from the reinforcing bars. The concrete remains in its carbonated state, but the pore water of the concrete has turned alkaline again and thus protects against corrosion.

Electro-osmosis

The alkaline electrolyte placed on the surface of the concrete will penetrate the concrete by capillary absorption (capillarity) and by electro-osmosis, which is electrolyte migration due to the difference in electric potential.

The alkaline electrolyte penetrates the surface layer of the concrete parallel with the concrete surface. After a suitable period of re-alkalization the penetrating electrolyte will meet the concentric zones formed around the reinforcing bars by electrolysis. Thus the re-alkalization is complete. However, it does not harm the reinforcement (or the concrete) that there are areas between the reinforcing bars which are not fully re-alkalized.

Surface protection

When satisfactory re-alkalization of the concrete is achieved, the concrete surface should be cleaned, e.g. by high-pressure washing.

If a solution with calcium hydroxide is used as electrolyte, the concrete surface should be protected by a carbonation retarding paint. Otherwise, the carbonation will continue, because calcium hydroxide will react with the carbon dioxide in the air resulting in continued carbonation. Therefore, it is necessary to protect the concrete surface.

If, however, a solution with soda is used as electrolyte, surface protection is not necessary. For a considerable period, soda will be transformed into sodium bicarbonate by penetration of CO_2 and the pore water in the concrete will have a pH value above 10.3. This will still protect the reinforcement from corrosion.

However, soda cannot always be used as electrolyte. If the concrete aggregates contain more than 2% porous flint (by volume), the alkali reaction will be facilitated as soda contains Na^+, which is an alkali metal ion.

There are many types of carbonation retarding surface protection (Annex E1). A suitable material may be one-component, aqueous co-polymer or cement-based, acrylic cement.

Practical experience

The widest experience with electrochemical re-alkalization can be found in Norway, UK, USA and Japan. Since 1988 electrochemical rehabilitation of approximately $250\,000\,m^2$ of concrete surface, and in 1995 alone $80\,000\,m^2$, has been performed. In Denmark there is positive experience with the few projects carried out (Poulsen, 1996; Ketelsen *et al.*, 2000).

Preliminary investigations

Prior to re-alkalization of a building component, investigations (registration and evaluation) of the concrete should be performed. There are two purposes of this investigation:

- Documentation of carbonation depth of the concrete. This knowledge should be used for planning the re-alkalization and for later determination of the efficiency of the alkalization.
- Demonstration of the damage and defects found in the concrete, e.g. initial rust on the reinforcement and delamination of the concrete. It is important for correct rehabilitation to determine whether the concrete suffers from other 'concrete diseases' than carbonation. If so, the rehabilitation should be adjusted accordingly.

It is very significant for successful rehabilitation to ascertain whether the concrete has damage other than that caused by carbonation. If the concrete shows weak signs of alkali–silica reaction or if the aggregate contains alkali-reactive particles, it may be significant to the choice of electrolyte.

Re-alkalization

When the result of the preliminary investigation is available the re-alkalization of the concrete can be started taking due account of the results of the investigation. Prior to re-alkalization it should be checked that the reinforcement of the concrete has full metallic contact so that the re-alkalization will be efficient. Quality assurance comprising at least the following should be performed:

- Continuous registration of voltage and amperage
- The electrolyte should be regularly checked and replenished when necessary
- Bored cores should be drawn once in the middle of the process. They should be analysed to establish how progress in the re-alkalization process will occur.

Documentation

When the continuous control testing of the concrete surface shows that the concrete has recovered its alkalinity, concrete cores should be drawn for final documentation of the re-alkalization. Decision as to where the concrete cores should be drawn as well as boring of cores should be made by an accredited concrete laboratory. The cores should form the basis of determination of the alkalinity of concrete by testing with phenolphthalein (colour indicator), pH profile and ion profiles, e.g. OH^- profile. The control testing should be performed on site and should be attended by the supervisor.

Based on measurement of the re-alkalization of the concrete surface it is possible to decide whether the re-alkalization has acted for a sufficiently long period.

The concrete should turn red over an area of at least 40 mm around the reinforcing bars when phenolphthalein is applied to the fractured surface of a bored core.

Need for re-alkalization in Denmark

Concrete facades are exposed to carbonation, especially facade components finished with a toe. At the back of the toe the concrete is protected from driving rain and therefore carbonates rather fast (faster than at front). With the w/c-ratio normally used for concrete in facade components, a carbonation depth of 20–30 mm over 20 years is not unusual. At the front of the toe the concrete carbonates slowly as the concrete is exposed to driving rain and flushing of water from the facade component.

The rust protection of the reinforcement will then disappear due to carbonation from the back and due to penetrating moisture, which drenches the concrete of the toe, so that there will be favourable conditions for corrosion resulting in bursting in the toe.

It is hardly necessary to re-alkalize the entire facade, but local re-alkalization of the toe is possible. The practical arrangement of the re-alkalization depends on the geometric shape of the toe and especially the tolerance.

Another possibility for rehabilitation of the facade is to cut off the toe and apply a filling, which will appear as a horizontal strip in the facade. However, this will not always comply with the architecture of the building.

Manufacturer/supplier: electrochemical re-alkalization

Method M7.3 only requires little special equipment besides that already available at manufacturers. This is:

- Anode grid, either titanium or steel grid (dependent on the task)

- Electric power unit to control the electric potential to be provided with anode and cathode (the reinforcement). It is preferable that this power unit is able to work with constant current density so that the area of the structure to be re-alkalized can be controlled by, for example, 1 A/m^2 concrete surface
- Various electric, insulated wires or cables. Wires to the anode (positive cables) should be red. Wires to the reinforcement (negative cables) should be black
- Electrolyte, e.g. calcium hydroxide or soda (sodium carbonate), dependent on the composition of the concrete
- Electrolyte container, e.g. papier mâché or textiles.

Designer/builder: electrochemical re-alkalization

Electrochemical re-alkalization of a concrete building component should fulfil certain requirements to be able to restore the passivity of the reinforcement without adverse effects. Therefore, in EN 14038, part 1, on 'Electrochemical re-alkalization', minimum requirements are specified for investigation of concrete and reinforcement before, during and after design of re-alkalization of a concrete structure.

Concrete investigations

There are three purposes of investigation of the concrete and reinforcement of the structure:

- To assess the physical state of the concrete structure with regard to codified safety and the necessary repair besides electrochemical re-alkalization.
- To decide whether or not a concrete structure is suitable for rehabilitation by electrochemical re-alkalization.
- To establish the necessary knowledge of the concrete structure and its present state to be used for planning of the electrochemical re-alkalization.

Assessment of existing drawings and description

The history of the concrete structure concerned should be elucidated, if possible. Here, existing drawings, description and previous investigations may contribute to the history of the structure. However, experience shows that this material is often imperfect or drawings and description have not been updated after construction. If this is the case, more details for the registration of the state of the concrete and reinforcement are required. In particular it is important to determine the type, location and quantity of the reinforcement. This should be verified by subsequent inspection.

Visual inspection

Further to verification of the information to be retrieved from existing drawings and description, the extent of damage to concrete and reinforcement should be registered. Thus, all areas where re-alkalization is planned should be checked for cracks, delamination (by, for example, impact echo), honeycombs and damaged construction joints. Furthermore, the thickness of the reinforcement cover should be registered.

In particular, all previous repairs should be investigated. When electrochemical re-alkalization is to be used for concrete behind such repairs, the resistivity and porosity of the repair concrete should be determined.

Representative Ø 75 mm concrete cores are bored out. These cores are subjected to micro-structural and chemical investigations, where the following should be determined:

- Carbonation depth according to EN 13295
- The w/c-ratio of the concrete
- The chloride ion content according to EN 13396
- Alkali reactive particles (porous flint) in the sand fraction.

Furthermore, all mutual metallic contact of the reinforcing bars should be documented and other metallic parts (inserts, etc.) should be put in conductive connection with the reinforcement (the cathode).

Adverse effects

Certain doubts have arisen with regard to adverse effects of electrochemical re-alkalization, for example, the question of alkali reaction and impairment of the anchorage capacity of the reinforcement by electrochemical re-alkalization.

Alkali reaction

If the concrete in the structure to be re-alkalized contains alkali reactive aggregate particles, development of alkali reaction should be prevented. In case of porous flint in the sand fraction, one of the following measures can be taken:

- If the content of porous flint in the sand is more than 2% (by volume), $Ca(OH)_2$ can be used as electrolyte. If so, continued carbonation of the concrete surface should be prevented by a carbonation retarding surface protection after re-alkalization.
- If the content of porous flint in the sand fraction is less than 2% (by volume), electrolyte determined by other conditions (e.g. economy) may be used.
- The above is particular to Danish situations with porous flint in the sand. In other countries where there are other forms of alkali reactive aggregate particles, other measures will, of course, have to be taken.

The first item can be clarified: If it can be established that a high-alkali cement is used (can be detected by thin-section analysis of the concrete) and the concrete has been placed in a humid outside environment without alkali reaction taking place, the risk of creating alkali reaction by electrochemical re-alkalization will be very low.

If it is detected that the concrete aggregate contains porous flint, specialist assistance is advised.

The significance of the electrolyte for development of an alkali reaction is explained (Ketelsen *et al.*, 2000). In this investigation electrolytes, such as solutions of saturated calcium hydroxide or saturated sodium carbonate (soda), were used as curing liquids for mortar prisms, where the sand fraction contained from 0 to 5% of porous flint (by volume) of particle size 2–4 mm. The period of exposure was 52 weeks at 50 °C.

From this mortar prism investigation it can be concluded that use of calcium hydroxide electrolyte does not result in crack inducing alkali reaction in flint-containing concrete by electrochemical re-alkalization. However, based on these mortar prism investigations, this is not the case when sodium carbonate is used as electrolyte. It has not been checked whether this conclusion is true in practice.

Mortar prism investigation is a comparative investigation of electrolytes. Only investigations on site can answer this question. The advantage using sodium carbonate as electrolyte is that:

- sodium carbonate is more efficient than calcium hydroxide as electrolyte. This means that the time of re-alkalization for sodium carbonate is much shorter (1–2 days)
- surface protection of re-alkalized concrete is not necessarily required if sodium carbonate is used as electrolyte.

In practice, however, surface treatment will almost always be necessary for aesthetic reasons.

Anchorage capacity of reinforcing bars

A significant property of reinforcing bars is their ability to be anchored in concrete. By cathodic action of the re-alkalization there is a drastic reaction on the surface of the reinforcement. This may reduce the anchorage capacity of the reinforcement.

To look into the hypothesis that the anchorage capacity of reinforcement is reduced by electrochemical re-alkalization, a laboratory investigation of Ø 10 mm embedded ribbed bars was carried out according to DS 2082 (Ketelsen *et al.*, 2000). Half of the test specimens were fully carbonated until formation of rust was established. Then all test specimens were subjected to electrochemical re-alkalization at different current densities (from 0 up to 1.2 A/m^2 of concrete surface corresponding to 2.4 A/m^2 reinforcement surface). It was determined that rust formation increases the anchorage capacity of the reinforcement (measured by the mean anchorage factor). It could not be demonstrated that the mean anchorage factor decreases with the current intensity by incipient rust formation on the reinforcement (due to high diffusion). However, it was established that for non-carbonated concrete the mean anchorage capacity of the reinforcement showed a gently declining tendency with the current intensity. It does not, however, get lower than the recommended value $\zeta = 0.6$ in DS 411:1999 when the re-alkalization fulfils the requirement of a recommended current density of 1 A/m^2 concrete surface. However, the characteristic anchorage factor declines somewhat more with the current density, but it does not get lower than the recommended value $\zeta = 0.6$ in DS 411:1999, regardless of the treatment.

The anchorage tests performed are not applicable to the conditions for plain reinforcement.

Contractor/supervisor: electrochemical re-alkalization

Execution of electrochemical re-alkalization of reinforced concrete structures should comply with certain minimum requirements to fulfil the requirements specified in EN 14038, part 1 on 'Electrochemical re-alkalization and chloride extraction treatments for reinforced concrete – Part 1: Re-alkalization'. This applies to documentation of the repair materials received, the execution of work and the finishing treatment.

Initial work

Prior to the re-alkalization process the concrete surface and its reinforcement should be prepared removing all that may reduce the efficiency of the re-alkalization.

Removal of concrete

Defects and similar, which may impair the electrochemical process, should be removed and the surface should be cleaned (i.e. cracks, delamination, spalling, honeycombs, areas with coarse porous concrete, metallic inserts, repairs with polymeric mortar and similar which may impede the efficiency of the re-alkalization). Where honeycombs or coarse porous concrete extend to the back of the principal reinforcement, at least 20 mm of concrete behind the reinforcing bar, which is located at the deepest level of the principal reinforcement, should be removed to ensure sufficient encasing of the reinforcement. It is not necessary to remove carbonated, but otherwise healthy (i.e. resounding), concrete before commencing electrochemical re-alkalization.

Preparation of reinforcement

Reinforcement with loose rust cake formation should be cleaned, e.g. by a rotating steel brush, to ensure good (electric) contact between the reinforcement and repair mortar. It is not necessary to clean down to bare metal (St 2 or Sa 2½). However, it is important not to damage the reinforcement and therefore, angle grinder should not be applied for cleaning (rotating steel brush or sand blasting is preferable).

Removal of reinforcing bars or supplement should be in accordance with the specifications of the rehabilitation project.

Concrete repair

Defects and damage should be repaired (Annex E6) observing the following:

- Primer, bond-improving agents or other surface protection should not be applied to reinforcement
- Bond-improving surface treatment should not be applied to cut surfaces
- Polymeric repair mortar (epoxy mortar) should not be applied. Repair mortars used should be comparable to the original mortar with regard to electric resistivity and porosity
- Repair mortar should be cement-based and fibres or powder of metal should not be used
- Finishing treatment of repairs by, for example, sealing, impregnation or paint should not be used before re-alkalization is started
- Any injection of cracks should be performed with injection liquid of cement slurry or cement mortar.

Materials and equipment

All equipment should be calibrated in accordance with national requirements or European standard for measurements.

Anodes and anode zones

The anode grid should be able to distribute the current as required in the project.

The concrete surface to be re-alkalized is divided into a number of anode zones. Each anode zone should ensure uniform distribution of the current density. An anode zone should typically not be larger than approximately $50\,m^2$ at a maximum current of approximately $100\,A$.

The anode grid is placed in electrolyte on the concrete surface and should be in contact with the surface. This electrolyte is carried by plastic containers applied to the concrete surface, or the concrete surface may be coated with papier mâché, textile or similar which should be saturated with the electrolyte. The electrolyte should be able to establish electric connection between anode grid and the reinforcement so that the carbonated concrete is re-alkalized.

All wires and cables should be insulated. Red wires should be used for the anode and black for the reinforcement.

Wires and cables should be dimensioned for the designed current plus 25% and should have sufficient mechanical strength.

Electric power supply

A power unit supplying alternating current and a transformer/unifier should be used for imposing the voltage. This power supply and transformer should fulfil the requirements of EN 60742 and, if necessary, be able to function at a temperature above 30 °C using oil-cooling or air-cooling. The equipment should be suitable for outdoor application.

The transformer/unifier should be able to maintain a constant electric current.

Installation of equipment

Mutual metallic contact of the reinforcing bars should be ensured before the equipment is installed (see the following section). Installation of the anode system should be in accordance with the requirements in national or European standards on electric installations.

Each anode zone to be re-alkalized should be provided with an adequate number of electric connections to the reinforcement.

The power supply of each anode zone should be registered with regard to consumption of ampere hours (A/h) and the time elapsed for re-alkalization.

Prior to installation of the anode system, the concrete surface should be checked for the possibility of short-circuits (e.g. honeycombs, embedded metal parts). Any areas of short-circuiting should be repaired by insulation, e.g. with silicone rubber or repair mortar.

The electrolyte should be protected from weathering (sun, rain, wind and frost). If papier mâché is used on the concrete surface it should be protected from gusts of wind, especially where the electrochemical re-alkalization takes place at high altitudes. The electrolyte should be protected from freezing. If the electrolyte freezes, the re-alkalization should be discontinued at once and should not be resumed before the electrolyte is fully defrosted. To ensure that the electrolyte does not freeze, heat blowers may be used. If the electrolyte dries out, it should be replenished.

Reinforcement continuity

The (electric) continuity of the reinforcement should be determined. The first information can be taken from the structural drawings of the structure concerned. There is no continuity between reinforcement in various prefabricated concrete components.

However, it may be expected that the reinforcement has continuity within the same concrete component, but this is not always the case. If it has been especially required that certain areas of the reinforcement should be insulated from other reinforcement, this will normally be specified in the original work specification.

Furthermore, reinforcement continuity (metallic connection) in site cast building components of concrete may be expected, but this should be verified by inspection.

Determination (measurement) showing that the reinforcement is (electrically) continuous (i.e. is in metallic contact) may be performed according to EN 12696, part 1: 'Cathodic protection of steel in concrete – Part 1. Atmospherically exposed concrete'. This investigation should comprise

- Continuity of reinforcement in all building components within a re-alkalization zone.
- Continuity of all metal parts (other than the reinforcement) to the reinforcement.

The principle of measuring continuity of reinforcement is to measure the resistance between reinforcement at different points of a re-alkalization zone. If the measured resistance exceeds $1\,\Omega$, it is highly probable that there is no metallic contact.

Commencement, operation and disassembling

When the equipment has been installed it should be checked directly before, during and after commencement of the electrochemical re-alkalization.

Investigations and documentation prior to commencement

Before commencement of the electrochemical re-alkalization, the following should be checked:

- The electric circuit should be checked
- The electric connections should be checked by measuring the electric resistance in all anode connections and all cathode connections
- Visual inspection that all wires and cables are correctly marked and that the equipment is properly protected from weathering and vandalism.

Testing and inspection during re-alkalization

At least once a day the current in wires to the anode zones to be re-alkalized should be measured. A current density of approximately $1\,A/m^2$ concrete surface for approximately 200 hours should be attained (or as specified in the work specification). The minimum re-alkalization time is 100 hours. However, it should be documented (by calculation) that the current intensity does not exceed $4\,A/m^2$ reinforcement surface. Furthermore, the following should be checked:

- Check that all re-alkalization zones function
- Check that all anode connections function
- Check that the electrolyte does not dry out and replenish if necessary.

Furthermore, the following shall be measured:

- The current (ampere, A) and the time elapsed (hours, h). This may be done electronically by an automatic Ah meter. The time intervals should not exceed 2 hours.
- Measure the voltage of each re-alkalization zone.
- Boring of concrete cores $\varnothing$ 50 for determination of re-alkalization after exposure of $200\,Ah/m^2$ reinforcement surface.

Determination of the 'carbonation depth' of the concrete surface

Distribution of the carbonation depth over the concrete surface by surface carbonation should be made according to EN 14630 on 'Products and systems for the protection and repair of concrete structures – Test methods: Determination of carbonation depth in hardened concrete by the phenolphthalein method' or similar test method. Surface carbonizing should be distinguished from crack carbonation. The carbonation depth should be measured on sawn or fresh fracture surfaces of core samples with a diameter of minimum Ø 50 mm. Sawn surfaces should be cleaned with pure, fresh water and thoroughly dried with a cloth directly before measuring of the carbonation depth by a pH indicator.

Use of universal pH indicator cannot be recommended. Tests have shown that such indicators can be very unreliable and give misleading results. Use of phenolphthalein as a pH indicator is normally recommendable. Usually there are no application problems before the electrochemical process when the test method in EN 14630 is observed.

On the other hand, there were certain problems with the colour transition during and after the electrochemical re-alkalization where the concrete has been saturated by electrolyte. The problems with use of phenolphthalein as a pH indicator on electrochemically re-alkalized concrete can be illustrated by the following observations. It is, for example, observed that for high pH values (close to 14 and above) red colour is not produced by application of phenolphthalein. Thus it can be observed that the red colour is either not produced or that it disappears quickly. This phenomenon can be observed in the concrete close to reinforcement, which has been treated by electrochemical re-alkalization (the cathode reaction takes place on the surface of the reinforcement, i.e. OH ions are produced).

Also, it has been observed that for low pH values (between 10 and 11) generation of the red colour depends on the composition of the indicator liquid (i.e. the concentration of phenolphthalein). By application of a pH indicator made by 1% phenolphthalein in 96% ethanol, the colour transition takes place at a pH value of 11. If a pH indicator is made by 0.5 g phenolphthalein of 500 ml ethanol, which is dissolved in up to 1 litre, the colour transition takes place at a pH value of 10. For pH values between 10 and 11 the red colour is, however, not as intense as for pH values above 11.

Use of phenolphthalein as a pH indicator requires a trained operator and in any case, use of phenolphthalein as a pH indicator can only be informative. The final documentation may also include determination of a pH profile determined by powdered concrete or, better, by determination of an OH^--profile, but this is normally not considered necessary.

Disassembling of equipment and final report

After checking that the re-alkalization is satisfactory, the anode equipment is disassembled. The concrete surface is cleaned by high-pressure washing and prepared for surface treatment, if any.

Final report

A report on the work and the measurements performed on the concrete structure should be worked out. This report should as a minimum contain the following information:

- Description of the concrete structure together with its location and physical condition
- Participants (and their responsibilities) in the electrochemical re-alkalization, i.e. consultant, contractor and subcontractors, supervisors and institutions (laboratories), who have made investigations of the concrete and reinforcement of the structure
- Description of the electrochemical re-alkalization performed
- Copy of the work specification with drawings (shall specify the present, rehabilitated state of the structure) and the applied acceptance criteria

- Description of the preparatory work prior to the electrochemical re-alkalization comprising preparation of the concrete surface and reinforcement, repair and sampling
- Collected control data. Registration of the ampere hours (Ah) used for each of the treated re-alkalization zones and the registered voltage potentials during the re-alkalization process
- Discussion and conclusion on the results obtained, inclusive of local deviations or variations in the re-alkalization process and evaluation of the achieved efficiency of the re-alkalization process
- Other relevant documents.

Instrumentation

It is possible to observe the state of the structure in a non-destructive way by mounting measuring equipment to monitor whether the effect of the electrochemical monitoring process is maintained, e.g. probes for registration of the corrosion state of the reinforcement (corrosion rate).

References

Poulsen. Betonrenovering med elektro-kemiske metoder. *Håndbog for Bygningsindustrien*, HFB 28. Nyt Nordisk Forlag Arnold Busk, København, 1996.

Ketelsen, Poulsen, Poulsen. Elektro-kemisk re-alkalisering af carbonatiserede betonfacader, en miljøvenlig, ikke-destruktiv og ressourcebesparende renoveringsform. Erhvervsfremme Styrelsen, Carl Bro Hillerød, 2000.

Poulsen *et al.* 13 betonsygdomme. Hvordan de opstår, forløber og forebygges. Statens Byggeforskningsinstitut, Beton 4, Hørsholm, 1985.

Annex E13: Electrochemical chloride extraction

Introduction

The standard EN 14038-2 on 'Electrochemical re-alkalization and chloride extraction for reinforced concrete – Part 2: Chloride extraction' will describe method M7.5 in ENV 1504-9 on 'General principles for application of products and systems'. The standard EN 14038-2 is in preparation, but publication is not expected in the near future.

Method M7.5 is applicable to electrochemical extraction of chloride ions from concrete where chloride penetration by diffusion has taken place. Method M7.5 is less suitable for extracting chloride mixed with concrete by the mixing process. Method M7.5 may be used for chloride-containing concrete with normal embedded reinforcement, which has been exposed to chloride penetration in a chloride-containing environment. However, method M7.5 is not applicable to chloride-containing concrete with prestressed reinforcement of high-strength steel or epoxy treated or galvanized reinforcement.

Background for method M7.5

Concrete in building components exposed to marine environment or road environment (parking decks) and concrete in some food industries, swimming baths and access balconies sometimes show damage caused by rust due to chloride attack.

Chloride should be removed before rust formation

Chlorides have caused a considerable amount of concrete damage all over the world. Only by periodic and careful inspection of concrete structures in chloride-containing environments, can incipient chloride attack be detected before the damage propagates to such an extent that rehabilitation will be so costly that replacement of the structure is considered instead of rehabilitation.

Once the chloride attack is in progress it will normally be very fast and the reinforcement may burst due to rust. Before this state the reinforcement will have lost its strength and ductility and the safety of the structure against rupture will no longer fulfil the codified requirements. However, it is possible to rehabilitate the concrete and save the structure before this stage.

Electrochemical extraction

Extraction of chlorides from the concrete is (theoretically) fairly simple. Chlorides are negatively charged ions. Concrete is porous and this porosity is more or less filled with moisture (pore water). The negatively charged chloride ions are therefore dissolved in the pore water of the concrete where they migrate (free chlorides). However, some chlorides are chemically and physically bound to the putty of the concrete.

By the electrochemical method the free (negative) ions are extracted from the concrete by using the fact that the negative chloride ions are repelled by (negative) voltage and attracted by positive voltage. Therefore, negative voltage is imposed on the reinforcement and wet textile, papier mâché or the like with a grid of titanium applied to the concrete surface and a positive voltage imposed. This is identical to electrochemical re-alkalization (Annex E12).

The result is that chlorides in the concrete are repelled by the negative reinforcement and attracted by the positively charged titanium grid on the concrete surface. A current of approximately 1 A per m^2 of concrete surface is supplied and in approximately 4–8 weeks the chlorides migrate from the concrete to the textile or the papier mâché which is placed on the concrete surface and contains the anode (positive).

When it is verified that the chloride extraction is satisfactory, the textile with the extracted chlorides is removed and the rehabilitation is finished apart from cleaning of the concrete surface and application of chloride-retarding surface treatment to prevent future chloride penetration. Chloride proof or chloride-retarding surface protection can only be omitted if the chloride action can be removed in other ways, e.g. by abandoning the use of chloride-containing road de-icers.

There are several chloride-proof and chloride-retarding surface protection agents on the market (Annex E1).

Practical experience

The widest experience with electrochemical chloride extraction can be found in Norway, UK, USA and Japan.

In Denmark a few projects have been carried out. The experience is positive provided intervention is made in time in case of chloride containing concrete, i.e. before the reinforcement has started corroding. If the reinforcement corrodes and this results in cracks, delamination and spalling of reinforcement cover, method M7.5 is more problematic. Concrete, which has delaminated due to chloride attack, should be removed and the surface should be repaired before the electrochemical chloride extraction is efficient in practice.

If the chlorides have been present in the concrete structure for such a period that large areas of the reinforcement are consumed by rust, not only repair and extraction of chlorides are required, but also strengthening of the concrete structure (Annexes E7, E8 and E9). Therefore, it may be very costly if chloride attack is detected too late.

The SHRP project

In the USA an extensive research and development programme on concrete has been carried out, named 'Strategic Highway Research Program' (SHRP). This is the greatest RD programme ever on rehabilitation of damaged concrete structures.

The SHRP included laboratory tests, as well as field investigations. The SHRP report of more than 200 pages with documentation of the applicability of the method in practice is valuable for those who want further information on electrochemical chloride extraction (Bennett *et al.*, 1992).

Preliminary investigations

Prior to commencement of chloride extraction, an investigation of the concrete should be made. There are three purposes of this investigation:

- Documentation of the depth of chloride penetration into the concrete. This knowledge should be used for planning the chloride extraction and for later determination of the efficiency of the chloride extraction.
- Determination of the chloride diffusivity of the concrete. The result may be significant to planning of the chloride extraction.
- Demonstration of the damage and defects found in the concrete, e.g. initial rust on the reinforcement and delamination of the concrete. It is important for correct rehabilitation to determine whether the concrete suffers from other 'concrete diseases'. If so, the rehabilitation should be adjusted accordingly.

It is very significant for successful rehabilitation to ascertain whether the concrete has damage other than that caused by chloride attack. If the concrete shows weak signs of alkali reaction, it may be significant to the choice of electrolyte.

Chloride extraction

When the result of the preliminary investigation is available, the electrochemical chloride extraction can be commenced. Prior to chloride extraction it should be checked that the reinforcing bars of the concrete are in full metallic contact with each other so that the chloride extraction will be efficient. This is in close accordance with electrochemical alkalization (Annex E12).

Quality assurance comprising at least the following should be performed during chloride extraction:

- Continuous registration of voltage and amperage
- The electrolyte around the anode on the concrete surface should be regularly checked and replenished when necessary.

Powder samples (or better, Ø 50 mm bored cores split and tested with chloride indicator) should be drawn regularly from the concrete surface. They should be analysed to establish the progress in the chloride extraction.

Documentation

When the continuous control testing of the concrete surface shows that the chloride content of the concrete has been reduced to a suitable level, Ø 75 concrete cores should be drawn for final documentation of the chloride extraction for determination of chloride profiles. These concrete cores are sent to an accredited laboratory for determination of the chloride profile of the concrete. Based on measurement of the chloride extraction achieved, it can be decided whether the chloride extraction has been active for a suitable period of time.

The goal is that the chloride content around the concrete reinforcement has reached a suitable level, i.e. 0.05 to 0.06% of the concrete mass after levelling when the chloride-retarding surface protection has been on the surface for a certain period of time. It is better, but more laborious, to measure and evaluate the remaining, free chlorides in the concrete.

Projects in Denmark

In other countries, chloride extraction has been performed on many buildings. Not many buildings in Denmark have been rehabilitated in this way. Rehabilitation of chloride-containing concrete by electrochemical chloride extraction has only been performed on the following structures in Denmark:

- Bridge pier for motorway bridge across Slotsherrensvej, Copenhagen county. Supervision: AEC.
- Swimming bath, Virum, Lyngby-Taarbaek municipality. Supervion: AEC.
- Concrete walls in tunnel under Bernstorffsvej, bridge section, Road Directorate. Supervision: Road Directorate.
- Balcony columns in residential houses, Vanløse, Lønstrupgaard. Supervision i-68.

Producer/supplier: electrochemical chloride extraction

Method M7.3 only requires little special equipment besides that already available at manufacturers. This is:

- Anode grid, either titanium or steel grid (dependent on task)
- Electric power unit to control the electric potential to be provided with anode and cathode (the reinforcement). It is preferable that this power unit is able to work with constant current density so that the area of the structure to be re-alkalized can be controlled by $1\,A/m^2$ concrete surface
- Various insulated electric wires or cables. Wires to the anode (positive cables) should be red. Wires to the reinforcement (negative cables) should be black
- Electrolyte, e.g. calcium hydroxide or soda (sodium carbonate), dependent on the composition of the concrete
- Electrolyte container, e.g. papier mâché or textiles.

Designer/builder: electrochemical chloride extraction

Electrochemical chloride extraction from a concrete building component should fulfil certain requirements to be able to restore the passivity of the reinforcement without adverse effects. Therefore, in EN 14038, part 1 on 'Electrochemical chloride extraction', minimum requirements will be specified for investigations of concrete and reinforcement before, during and after design of re-alkalization of a concrete structure. In principle, method 7.5 is no different from method M7.3, but method M7.3 requires a longer time (from 4 to 8 weeks). Until EN 14038 part 2 is available, EN 14038 part 1 could apply.

Contractor/supervisor: electrochemical chloride extraction

Electrochemical chloride extraction from reinforced concrete structures should fulfil certain minimum requirements to be in accordance with EN 14038, part 2 on 'Electrochemical re-alkalization and chloride extraction treatments for reinforced concrete – Part 2: Chloride extraction', when this standard is available. The requirements will include documentation of the repair materials, the works and the finishing treatment.

In principle, method 7.5 is no different from method M7.3, but method M7.3 requires a longer time (from 4 to 8 weeks). Until EN 14038 part 2 is available, EN 14038 part 1 could apply.

References

Comprehensive technical literature on the subject is available, among others

Bennett *et al*. Evaluation of Norcure process for electrochemical chloride removal from steel reinforced concrete bridge components. SHRP-C/URF, Washington, DC, 1992.

Poulsen *et al*. Elektro-kemisk chloridudtrækning af beton. Dansk Beton nr. 3, København, 1988.

Sørensen. Chloridudtrækning, nye forskningsresultater – Et alternativ til katodisk beskyttelse. Dansk Selskab for Materialeprøvning og -forskning. København, 1993.

Sørensen. Elektrokemiske reparationsmetoders anvendelse i forbindelse med korrosion af armering i beton. Erhvervsforskningsprojekt, PhD, EF307. ATV, COWI Lyngby, 1995.

Annex E14: Cathodic protection of reinforcement

Introduction

The standard EN 1504-7 on 'Corrosion protection: coatings for reinforcement' specifies requirements for surface protection of reinforcement in concrete, i.e. principle P11 (on control of the anode areas of the reinforcement). This is explained in Annex E15. Annex 14 deals with principle P10: (method M10.1: Cathodic protection) using the standard EN 12696-1 on 'Cathodic protection of steel in concrete – Part 1: Atmospherically exposed concrete'. So far, there is no standard on cathodic protection of reinforcement in concrete subjected to aggressive environments, e.g. marine structures, water tanks and swimming baths.

Design, execution and supervision of cathodic protection is specialist work. Therefore, it is not the purpose of this annex to give guidance on the design and execution of cathodic protection of concrete reinforcement, but to provide understanding of the principles of cathodic protection and inform on the advantages and drawbacks of method M10.1 based on the standard EN 12696.

Background for method 10.1

As previously described (in Section 6.2.2, P10, Cathodic protection of concrete), cathodic protection can be established in the following way:

- M10.1a: *Passive cathodic protection*. The reinforcement of the building component is connected (electrically) with a sacrificial anode of, for example, magnesium, aluminium or zinc placed in an electrolyte in contact with the reinforcement.
- M10.1b: *Active cathodic protection*. The reinforcement of the building component is given a negative electric field with external source of current, and an anode (e.g. in the form of titanium grid, electrically conducting paint or an embedded anode, e.g. titanium bars) is placed on the concrete surface.

Inspired by Faraday, passive cathodic protection of steel in aggressive exposure class was invented in 1924 by Humphry Davy for use in the British Navy. At the beginning of the twentieth century cathodic protection was used for underground pipelines of steel. As it was detected that in certain areas the electrical resistance in the ground could be too large to perform passive cathodic protection, an electric field was imposed. Thus, cathodic protection was invented.

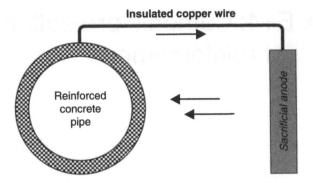

Ground surface

Figure E14.1 Example of passive cathodic protection of a reinforced concrete pipe in moist and aggressive soil, e.g. chloride containing, elevated seabed. A sacrificial anode is placed along the pipeline to ensure uniform distribution of the cathodic protection. The sacrificial anode is submerged and backfill material is applied, for example, betonite or cinder to ensure good electric contact between sacrificial anode and the ambient, moist soil (see Fontana, 1967).

However, it was not until many years later that cathodic protection for reinforced concrete structures in aggressive exposure class was developed. One of the earliest applications of cathodic protection of reinforcement in concrete was around 1950, where concrete tubes wrapped with prestressed reinforcement were given cathodic protection in the USA (Gourley, 1976; Heuze, 1965). Cathodic protection of bridge decks in California was the first application of cathodic protection of traditionally reinforced concrete. It was done in 1968 and since then, cathodic protection has been applied to a large number of reinforced concrete structures in the USA, UK, Hong Kong and the Middle East. Today it may be assumed that cathodic protection of approximately 1 mio. m^2 per year is performed. In Denmark cathodic protection of reinforcement has been used since 1970 (Arup, personal communication), among others for swimming baths, car parks, access balconies and bridge structures.

In concrete, which is free of chlorides and with a high pH value, the reinforcement will normally be passive. In carbonated concrete or concrete where chloride from the environment has penetrated, the reinforcement will normally corrode (Figure E14.2).

The principle of cathodic protection of reinforcement in concrete is to force the entire reinforcement to act as a cathode.

Passive cathodic protection of reinforcement in concrete

Passive cathodic protection of reinforcement is performed by connecting the reinforcement (with metallic contact) to a sacrificial anode. This sacrificial anode should consist of a metal, e.g. zinc, aluminium, magnesium or their alloys, which is less precious than the steel reinforcement and therefore will have more tendency to corrode than the reinforcement. The sacrificial anode corrodes and thus protects the reinforcement against corrosion. With this arrangement the reinforcement of the concrete, which is in contact with the sacrificial anode, becomes negative so that the reinforcement acts as a cathode. The sacrificial anode is placed in a moist environment, e.g. soil or seawater. It is significant that all reinforcing bars

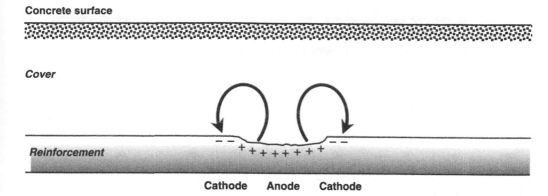

Concrete surface

Cover

Reinforcement

Cathode   Anode   Cathode

Figure E14.2   Schematic representation of the corrosion of a reinforcing bar. At the anode (+) the steel is dissolved and electrons are liberated so that corrosion flows towards the cathodes (−) (see Mays, 1992).

have mutual electric contact and that the electric resistivity between sacrificial anode and reinforcement through, for example, soil and concrete is not too high (Figure E14.1).

Passive cathodic protection of reinforcement may be used for prestressed concrete in seawater since the risk of formation of hydrogen brittleness can be ignored and since the electric resistance from sacrificial anode to reinforcement through seawater and (wet) concrete is low (Kessler *et al.*, 1995).

One of the advantages of passive cathodic protection (where possible) is that the method is independent of external power supply, which may be difficult and costly to produce for some concrete structures, e.g. in the wilderness.

A special design of passive cathodic protection is application of a zinc plate embedded in high-alkaline mortar as sacrificial anode (Galvashield anode). The zinc plate is connected to reinforcement to protect it against formation of incipient anodes. The zinc plate has to be covered with a repair mortar.

Active cathodic protection of reinforcement in concrete

Different forms of anodes can be applied to the concrete surface (Figure E14.3). The most general types are:

- Grid or similar of titanium, embedded in a mortar layer on the concrete surface
- Inserted anodes of titanium surrounded by a conductive paste
- Strips of titanium or carbon fibre embedded in milled grooves in the concrete surface.

Types that are not so generally applied today are:

- Electrically conductive mortar (containing carbon fibres) applied to the concrete surface
- Electrically conductive, organic paint on the concrete surface
- Electrically conductive, metallic paint on the concrete surface.

If a positive voltage is applied to the anode (by an external power supply) and a negative voltage is applied to the reinforcement of the concrete, cathodic protection of the reinforcement will be achieved. For normal, active cathodic protection the power consumption will typically be approximately 2–20 mA/m^2 of the steel reinforcement at a voltage of approximately 2–6 volts.

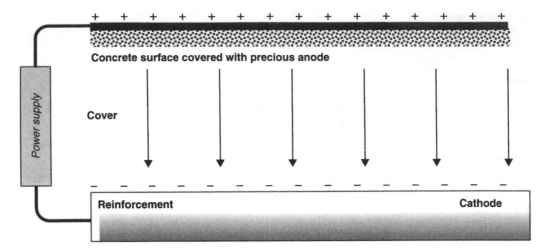

Figure E14.3 Schematic representation of cathodic protection. A precious anode in the form of electric paint, electrically conductive mortar or titanium grid in mortar coating is applied to the concrete surface. The sacrificial anode is connected electrically to the reinforcement via a rectifier and an electric field is imposed so that the reinforcement becomes negative and the surface anode becomes positive. Since the reinforcement becomes negative all over, it will not corrode and since the sacrificial anode is precious, nor will it corrode (Mays, 1992). Due to the electric field, ions will be transported in the reinforcement cover. Thus, chloride ions Cl^- will move away from the reinforcement and towards the anode. Similarly, sodium ions Na^+, potassium ions K^+ and calcium ions Ca^+ will be attracted by the reinforcement. At the anode, hydroxide ions will be consumed and oxygen and perhaps free chlorine will be produced. In the long term, acidification at the anode may take place which will dissolve the cement paste and thus reduce the bonding of the protective mortar layer. At the cathode, hydroxide ions OH^-, which build a protective passive layer on the reinforcement, will be produced. Too high current density may cause formation of harmful hydrogen concentrations H_2 at the reinforcement.

Producer/supplier: cathodic protection of reinforcement

Cathodic protection of reinforcement in a reinforced concrete structure is a system, which requires some equipment where the anode is the critical part. The standard EN 12696, part 1 on 'Cathodic protection of steel in concrete – Part 1: Atmospherically exposed concrete' gives different suggestions for anodes. Further, other equipment in the form of instrumentation of the reinforcement, power supplies and cables is necessary. Cathodic protection is applied as a system which can be adapted to the concrete structure to be rehabilitated.

Builder/designer: cathodic protection of reinforcement

First the builder/designer should assess whether the reinforced concrete structure concerned is suitable for cathodic protection. A reinforced concrete structure should possess certain properties to be suitable for cathodic protection of the reinforcement. Thus, in the standard EN 12696, part 1 on 'Cathodic protection of steel in concrete – Part 1: Atmospherically exposed concrete' the following requirements are made:

- All available drawings and descriptions for the structure concerned should be carefully studied to determine the location, quantity, dimension, type and electric continuity of the reinforcement and the composition of the concrete. This information should be checked later when the state of the structure is registered.

- For registration of state of concrete structures it is especially important to determine defects such as cracks, honeycombs and delamination. Also, old repairs and injections should be identified, if possible, because repair with polymeric repair materials and injection may have an insulating effect and thus reduce the efficiency of the cathodic protection.
- The chloride penetration and the carbonation depth should be determined and registered.
- The thickness of the reinforcement cover should be measured and registered since the electric resistance is significant to the choice of anode system.
- The electric continuity of the reinforcement should be checked, and the following should be fulfilled within the zone to receive cathodic protection:
  - Electric continuity between reinforcement in the different building components of the structure.
  - Electric continuity between reinforcing bars within the individual building components.
  - Further, other metallic installations, if any, in the concrete structure and their electric continuity with the reinforcement should be checked.
- Measurement of the electrochemical potential of the reinforcement (ekpm) with necessary break-ups to map the corrosion condition and the conductivity of the concrete. The state of reinforcement corrosion can be registered, e.g. in a grid with mesh width of 500 mm.
- All previous repairs should be removed if their electric resistivity is not in the range between 50 and 200% of the resistivity of the original substrate (e.g. a resistivity of 10–50 kΩcm).
- Metallic inserts as nails, wires and fittings that may come into contact with the anode, should be assessed and earthed or removed if necessary and the concrete should be repaired.
- Chloride containing, but otherwise healthy, concrete need not be removed.
- Loose rust on corroding reinforcement should be removed, but the reinforcement need not be cleaned down to the metallic exposed surface. Electrically insulating primer and rust protection should not be applied to the reinforcement.
- Where the concrete surface is repaired, the electric resistivity of the repair mortar should be in the range 50 to 200% of the resistivity of the original substrate. All repair mortar should be cement-based (but polymer-modified repair mortar is acceptable). Repair mortar with steel fibres should not be used. Curing membrane should not be used.
- The average pull-off strength of repairs should be more than 1.5 MPa and all strength observations should exceed 1.0 MPa unless the substrate fails.
- A pilot test to measure the effect of the cathodic protection should be made, so that it is possible to form an impression of the applicability of the method.

Considerations in connection with planning of cathodic protection

There are pros and cons by cathodic protection of reinforcement in concrete structures which should be evaluated in each individual case. The advantages are:

- Cathodic protection is time-saving compared to many other methods of rehabilitation. Often the structure can function as usual during application of cathodic protection.
- Cathodic protection should be used before significant damage arises on the concrete structure.
- Use of cathodic protection may be economical compared to other methods of rehabilitation. For example, chloride containing, but otherwise sound concrete, need not be removed.

- Bracing of the concrete structure during installation of cathodic protection may be less required than for other rehabilitation methods, since cathodic protection does not require removal of chloride containing, but otherwise sound concrete, for example.

The drawbacks to be considered when using cathodic protection are:

- The residual load-carrying capacity of concrete structures should fulfil the codified requirements after the corrosion of the reinforcement has been stopped. If not, the structure should be strengthened (e.g. supplementary reinforcement should be embedded or carbon fibre strips should be applied, see principle P4).
- Cathodic protection is solely used to stop reinforcement corrosion. If, moreover, there is concrete damage, it should be repaired by using principles P1 to P6. Rehabilitation of concrete damage in concrete structures may require intervention in the concrete of the structure so that it might be economical to rehabilitate the corrosion damage by methods other than cathodic protection.
- To apply cathodic protection, all reinforcing bars should be electrically connected (i.e. in metallic contact). Normally this is not a problem, but it may be necessary to create electric contact by welding, for example.
- Cathodic protection of prestressed concrete may be a problem, as in the previous explanation.
- Cathodic protection of reinforcement in concrete structures requires monitoring.
- The lifetime of the different anode systems varies from approximately 10 to more than 20 years (Broomfield, 1997).
- Cathodic protection systems require continuous supervision throughout the residual lifetime of the structure.

Adverse effects

Cathodic protection of concrete structures may, according to the literature, have different adverse effects. The builder/designer should evaluate the significance of these adverse effects before deciding on cathodic protection (DBF, 1992; Rilem, 1994).

Stray current

When a buried steel tank should have cathodic protection, for example, this is performed by burying an anode close to the tank. If there are other structures (e.g. a reinforced concrete pipe or reinforced concrete sheet piling) in the neighbourhood, they will also be subject to electric action (Figure E14.4). This means that local corrosion may take place in the concrete pipe due to stray current, unless the entire system of tank and pipeline is protected as a unit.

In connection with active earth wires (trams, trains, etc.), stray current resulting in corrosion on adjacent reinforced concrete structures may arise. If a reinforcing bar is thus given a positive potential, the bar will corrode where the current leaves the bar. However, this is a side effect of cathodic protection.

Hydrogen brittleness

Many standards on cathodic protection (e.g. DBF, 1992; Rilem, 1994; EN 12696, part 1) claim that cathodic protection should not be applied to prestressed concrete without further investigation, because prestressed reinforcement may develop hydrogen brittleness with risk

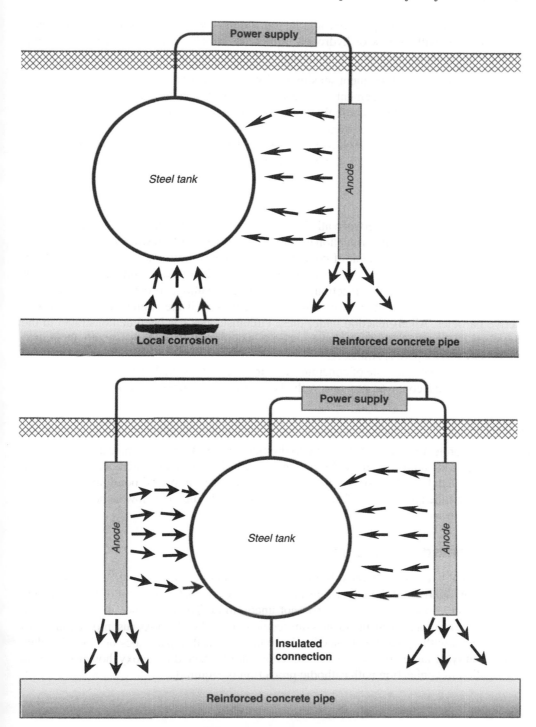

Figure E14.4 Top: Active cathodic protection of a steel tank in moist, aggressive soil may result in local corrosion on an adjacent reinforced concrete pipe (Fontana, 1967). Bottom: This may be prevented by placing anodes as shown (Fontana, 1967), and supplement with an electric connection to the reinforcement in the concrete pipe.

of sudden failure. However, passive cathodic protection may be applied for very humid concrete structures, without risk of hydrogen brittleness (Kessler *et al.*, 1995). Before taking this step, the following should be considered:

- Is the prestressed reinforcement sensitive to hydrogen brittleness?
- Is the prestressed reinforcement located in a steel pipe or behind a (Faraday's) cage of steel reinforcement so that cathodic protection will have no effect?
- Is it possible to ensure by instrumentation of the prestressed reinforcement that hydrogen brittleness cannot develop? Hydrogen is developed at a negative electric potential exceeding $-900\,mV$, ref. Ag/AgCl (i.e. with numerically larger value) (EN 12696-1).

Alkali reaction

Alkali reaction takes place between alkali reactive particles in the aggregate in humid, alkaline environments. Cathodic protection of reinforcement takes place when a negative charge is imposed on the reinforcement of concrete and attracts sodium and potassium ions from the pore water of the concrete. At the same time, hydroxide ions (alkaline) are developed at the reinforcement (the cathode reaction). Thus, in case of cathodic protection, the conditions for alkali reaction around the reinforcement will be present if the aggregate contains alkali reactive particles. Alkali reaction under these circumstances has been demonstrated in the laboratory. However, alkali reaction has not been established in practice by cathodic protection, cf. EN 12696, part 1, Annex A5.

In a registration of the state of concrete structures it is established that cathodic protection does *not* occur in a concrete structure with alkali reaction (Bennett, 1993).

Acidification

By cathodic protection of reinforcement, hydroxide ions are formed by the cathode reaction. This increases the tendency that chloride ions are repelled from the concrete reinforcement. On the other hand, acidification at the anode may result in impaired bonding of the cover which protects the anode. Therefore, the anode current should be as low as possible so that acidification is avoided.

Epoxy-coated reinforcement

In Florida, USA, epoxy is applied to many marine structures. In Denmark no such protection is made (apart from the Great Belt tunnel). However, rust protection of reinforcement with epoxy in connection with repair is not uncommon. The difference between original epoxy-coated reinforcement and local reinforcement treated with epoxy rust protection is that reinforcement with epoxy rust protection in connection with repair, does not lose the electric continuity. If large areas of the reinforcement are protected by epoxy, however, special investigation in connection with cathodic protection is required.

Design

Design of cathodic protection should always be performed by a group of technically skilled persons where at least one of them has theoretical and practical experience from previous projects. Design of protection is still more 'art' than 'science'. Therefore, the practical experience is a solid basis for design of cathodic protection.

Choice of anode system

If cathodic protection has been chosen as the method of rehabilitation, the anode system should be decided upon. For reinforced concrete structures exposed to the atmosphere, active cathodic protection is always chosen. For reinforced concrete structures with suitable humidity in marine environment, passive cathodic protection may be chosen. For the choice of anode system, the following guidelines can be used. If the concrete surface is subjected to traffic or otherwise exposed to abrasion, electrically conductive paint is not applicable as the anode system.

There has been considerable success with the following systems:

- Titanium grid protected by mortar cover, durability minimum 20 years.
- Strips of titanium or carbon fibre in milled grooves in the concrete surface. The technique with milled grooves is particularly applicable where there are restrictions with regard to the height or the self-weight of the structure. The space between the milled grooves should be approximately 200 mm (to be measured) and it is a laborious process. Therefore, the system is limited to application where other methods cannot be used. The durability is a minimum of 20 years (Broomfield, 1997).
- Anodes as inserts for cathodic protection of reinforcement from local corrosion.

Alternatively, and today more rarely, systems have been tentatively applied:

- Electrically conductive mortar (made conductive by addition of, for example, carbon fibres).
- Electrically conductive paint or metal spray (e.g. zinc). The advantages are that:
  - Paint does not increase the self-weight of the building component very much.
  - Paint is easily applicable to building components of complicated geometry.
  - Paint is very cheap and simple to repair (Broomfield, 1997).

In return, paint is not very resistant to mechanical action, e.g. abrasion, and the lifetime is limited to approximately 10 years (Broomfield, 1997). However, use of electrically conductive material may increase the resistance of the concrete surface to water vapour diffusion and thus the possibility of moisture damage.

Choice of electric potential

After the choice of anode system, the necessary electric potential should be determined to stop the corrosion. However, it is necessary to distinguish between corrosion caused by carbonation and corrosion caused by chloride ions (EN 12696, part 1).

Corrosion due to carbonation (Pourbaix diagram)

The state of reinforcement in (chloride-free) concrete depends on the pH value of the concrete (close to the reinforcement) and the electric potential of the reinforcement. This was first described by Pourbaix (1962, 1973) (Figure E14.5). Thus there are conditions where steel in concrete corrodes and conditions, where the reinforcement is passive. As mentioned, this depends on the pH value of the concrete and the electric potential of the reinforcement, but there are other significant parameters, e.g. the relative humidity of the concrete.

In the design of cathodic protection the issue is to change the electric potential of the reinforcement so that a stable condition without corrosion is achieved (i.e. the immune zone).

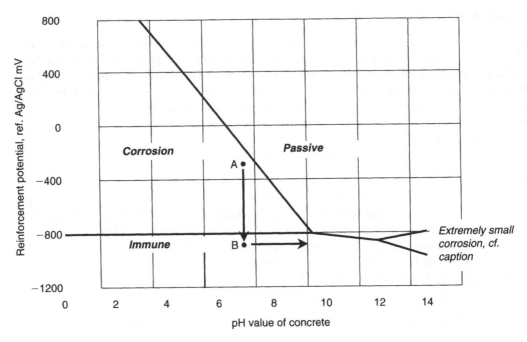

Figure E14.5   Diagram of theoretical pH potential for steel in chloride ion free water (Fe–H_2O system) at 25 °C (Pourbaix, 1962). Other parameters, e.g. temperature, are not taken into account. The steel potential should not be below −900 mV Ag/AgCl if the steel is sensitive to hydrogen brittleness. The Pourbaix diagram does not show the corrosion rate. Thus, the corrosion rate is extremely small in the triangular corrosion area in the Pourbaix diagram for $12 \leqslant pH \leqslant 14$. If a reinforcement is characterized by point A, corrosion takes place. By lowering the potential to B, the corrosion stops and the cathode reaction starts, i.e. point B moves to the right.

The electric potential for reinforcement of high-strength steel should, however, not be more negative than −900 mV, ref. Ag/AgCl (EN 12696). For a potential below −1100 mV, hydrogen may develop at the cathode. By the development of hydrogen the bond between the reinforcement and concrete is reduced, and certain steel types (e.g. prestressed reinforcement) may develop hydrogen brittleness.

However, the potential of the reinforcement to the immune zone will not be reduced. When cathodic protection is established on chloride free or carbonated concrete, hydroxide will develop at the reinforcement (cathode reaction) and the pH value will rise. Unless the chloride content of the concrete is very high, it will be possible to reduce the potential to the passive zone.

The Pourbaix diagram in Figure E14.5 is simplified and depends on parameters other than those specified here. Therefore, there are the following limits to the Pourbaix diagram (Berkeley and Pathmanaban, 1990):

- It is assumed that the pH value of the surface of the reinforcement is the same as in the pore water of the concrete. This is not always the case.
- The Pourbaix diagram is determined at 25 °C. As the corrosion process of the steel reinforcement is dependent on temperature, this may lead to misinterpretation.
- It is assumed that the concrete is free of chloride ions. However, this is not always the case in practice and should be taken into consideration (Figure E14.6).

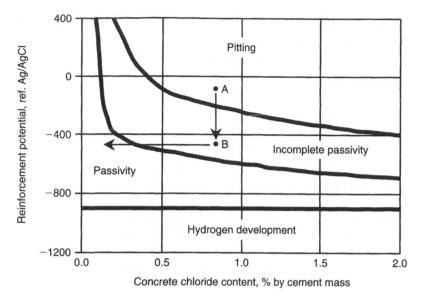

Figure E14.6   Theoretical and simplified chloride potential diagram for steel in humid, non-carbonated concrete at 25 °C with supply of oxygen (EN 12696, part 1, annex A, figure A.3). Other parameters such as temperature and moisture content of the concrete are not taken into account. Corrosion rate is not shown in the diagram. Explanation: pitting, pit corrosion is developed and propagates; incomplete passivity, pit corrosion is not developed, but already formed pitting is increased. If a reinforcement is characterized by point A, corrosion takes place. By lowering the potential to B, the corrosion does not stop, but the cathode reaction starts and chloride ions migrate away from the reinforcement, i.e. point B moves to the left and corrosion stops.

- The corrosion rate is not constant in corrosive areas. For example, the corrosion rate is extremely small in the small triangular corrosion area in the Pourbaix diagram, i.e. the area $12 \leqslant pH \leqslant 14$.

Corrosion due to chloride attack

The state of reinforcement in (non-carbonated concrete) concrete depends on the chloride content of the concrete (close to the reinforcement) and the electric potential of the reinforcement (EN 12696, annex A2, figure 6).

The diagram in Figure E14.6 is simplified and depends on parameters other than those specified here. Therefore, there are the following limits:

- It is assumed that the chloride content of the surface of the reinforcement is the same as in the pore water of the concrete. After application of the cathodic protection, the chloride ions migrate away from the reinforcement. This means that the electric potential of the reinforcement should be changed to maintain constant corrosion protection.
- The diagram is determined at 25 °C. As the corrosion process of the reinforcement is dependent on temperature, this may lead to misinterpretation.
- It is assumed that the concrete is not carbonated. However, this is not always the case in practice and should be taken into consideration.

Contractor/supervisor: cathodic protection of reinforcement

Cathodic protection should always be performed by skilled personnel, where at least one person has practical experience from previous projects with cathodic protection of reinforcement in concrete structures. It is important that quality control during the execution is well planned and described and that control is always made by persons who are specially instructed for performing this task.

Electric continuity of reinforcement

It is imperative that all reinforcing bars have mutual, mechanical contact or that it can be established, e.g. by applying supplementary reinforcing bars by welding or by connecting them so that mutual electric continuity is established.

Based on Danish experience from several measured structures, the requirements of EN 12696-1 for electric continuity ($1.0\,\Omega$ for $<1\,mV$) and for maximum resistance of cable/reinforcement ($0.01\,\Omega$) are not acceptable values. It is not recommended to use resistance measurement in case of direct current to control the reinforcement continuity in existing structures, because such measurement cannot verify that the reinforcing bars have mutual 'metallic' contact.

Reinforcement connections

Connection to the reinforcement should be performed with bolted or welded connections or shot-fired nails. If bolted joints include other metals (e.g. copper), embedded connection points should be sealed with dense mastics to prevent galvanic corrosion until the cathodic protection can be put into operation.

The number and location of connections should be chosen in connection with detailed design of the installation.

Power supplies

The power supply to the anode sections should be direct current. The regulation may be controlled by voltage (typically in the range 0–15 volts), controlled by power supply with individual exits to the single anode sections or in special cases controlled by potentials. The control boards with power supply and other electronics should be built in accordance with the rules valid for the application concerned.

Repair

The concrete structure to be repaired should be carefully checked and all previous repairs and existing damage in the form of defects and delamination should be assessed. To achieve approximately uniform conductivity, the former polymer-bound repairs should be replaced and new repairs of defects and delamination should be made with cement-bound repair mortar. Additives such as micro-silica and bond improving of polymers should not be used, since they will reduce the conductivity. Repair with cement-based sprayed mortar is applicable.

The anode system

The anode system specified in the project should be installed with careful observation of the specifications of the manufacturer. In particular, it should be controlled that embedment mortar and the workmanship ensure good electrolytic contact to the substrate.

Reinforcement and power supply

Connection to the reinforcement should be made with strong bolted connections or similar. The power supply should be direct current and be able to supply a voltage of 2–5 volts.

Monitoring equipment

Monitoring is very important to document proper performance. To ensure that the cathodic protection performs satisfactorily, monitoring equipment should therefore be installed. There are various forms of equipment in all price ranges.

Operation and maintenance

In connection with completion of a plant, a service and maintenance manual should be drawn up. This should among other things contain drawings and diagrams of the system with information of the individual components and documentation of configuration of the plant.

Cathodic protection of reinforcement in concrete structures requires continuous monitoring in the form of inspection and maintenance. It should be checked that the system performs as intended. Therefore, cathodic protection of reinforcement in concrete structures in the wilderness is less appropriate. However, cathodic protection of reinforcement in concrete structures already being regularly inspected is suitable. This applies to swimming pools, for example. Inspection (e.g. once a year) should normally comprise the following:

- The concrete structure should be visually inspected and the electric equipment and embedded reference cells should be checked.
- Measuring of the potentials should be continuously registered to assess whether the protective current has the correct value.

References

Alekseev, Ivanov, Modry, Schiessel. *Durability of reinforced concrete in aggressive media.* Balkema Publishers, Brookfield, VT, USA. Russian Translation Series no. 96, 1990.

Bentur, Diamond, Berke. *Steel corrosion in concrete, fundamentals and civil engineering practice.* E&FN Spon, London, UK, 1997.

Berkeley, Pathmanaban. *Cathodic protection of reinforced steel in concrete.* Butterworths, London, UK, 1990.

Broomfield. *Corrosion of steel in concrete, understanding, investigation and repair.* E&FN Spon, London, UK, 1997.

Chess. *Cathodic protection of steel in concrete.* E&FN Spon, London, UK, 1998.

Dansk Svømmebads-Teknisk Forening. Katodisk beskyttelse af svømmebassiner. Publikation nr. 52/1999.

DBF. Dansk Betonforenings Anvisning i Katodisk Beskyttelse. Dansk Betonforening Publikation 38. København, 1992.

Fontana. *Corrosion engineering.* McGraw-Hill, New York, USA, 1967.

Gourley. Practical results from the cathodic protection of steel in concrete. *Rocla Tech Journal*, December 1–8, 1976.

Heuze. Cathodic protection of steel in prestressed concrete. *Materials Protection*, 11: 57–62, 1965.

Mailvaganam. *Repair and protection of concrete structures*. Chapter 12: Corrosion protection. CRC Press, Boca Raton, FL, USA, 1991.

Mays. *Durability of concrete structures, investigation, repair, protection*, 1992.

Berkeley. Chapter 5: Protection. E&FN Spon, London, UK, 1997.

Pourbaix. *Atlas of potential/pH diagrams*. Pergamon Press, Oxford, UK, 1962.

Pourbaix. *Lectures on electrochemical corrosion*. Plenum Press, New York, USA, 1973.

prEN 12696 part 1. Cathodic protection of steel in concrete – Part 1: Atmospherically exposed concrete. CEN, Brussels, Belgium, August, 1999.

Rilem. Draft recommendation for repair strategies for concrete structures damaged by steel corrosion. Rilem Committee 124-SRC. Materials and Structures, 27, 1994.

Sandberg. Chloride initiated reinforcement corrosion in marine concrete. Lund University, Lund Institute of Technology, Division of Building Materials. Report TVBM-1015. Lund, Sweden, 1998.

Annex E15: Corrosion protection of reinforcement

Introduction

The standard ENV 1504-9 on 'General principles for the use of products and systems' describes the principle P11 on 'Control of anode areas of the reinforcement' and gives a number of examples:

- Method M11.1: Applying reinforcement cover with active pigments
- Method M11.2: Applying electrically insulating cover to reinforcement
- Method M11.3: Application of corrosion inhibitors for repair (Annex E16).

In EN 1504-7 on 'Corrosion protection: coatings for reinforcement' specific requirements for materials for surface treatment of reinforcement (corrosion protection) are made. It should be noted that the standard draft of EN 1504-7 has yet to be completed, but it is evaluated that the main content of the standard will be representative of the final standard.

Background for methods M11.1 and M11.2

An active corrosion on reinforcement will produce an electric current, the so-called corrosion current. This current flows between anode and cathode, and the anode area corrodes. There are two possibilities for stopping active corrosion:

- The anode area can be provided with a cover with pigments (of zinc, for example) which will corrode, but the reinforcement will not corrode (sacrificial anode) (method M11.1).
- The anode area can be insulated so that the corrosion current stops. However, it should be checked that a new anode area is not developed (incipient anodes) (method M11.2).

Corrosion protection of reinforcement

In the course of time, different materials (paints) for corrosion protection of reinforcement have been developed:

- Cement grout with addition of polymeric or latex emulsion and possibly corrosion inhibitor
- Solvent-free epoxy paints (EP)
- Polyvinyl chloride paints (PVC)
- Sacrificial paints, e.g. paint rich in zinc.

There is a lack of standards (besides EN 1504-7) for corrosion protection of reinforcement, but there is an instruction, e.g. BPS 122 (1997), and the manufacturer's instructions.

Method M11.1

The active pigments of the anode coating of the reinforcement may act as anodic corrosion inhibitors or as sacrificial anodes (e.g. paint rich in zinc and hot-dip galvanizing). Corrosion inhibitors are chemical additives which counteract formation of anode areas on the reinforcement. There is still some discussion on the long-term effect of corrosion inhibitors.

Application of reinforcement coating (paint) containing pigments of lower electric potential than reinforcing steel will have the effect that the cations of the cover are dissolved in the corrosion process instead of the steel of the reinforcement and thus protecting the reinforcement against corrosion. The metal ions of the cover may also become active in case of random (mechanical) damage to the reinforcement where the steel is exposed.

Other methods include phosphor with dissolved phosphoric acid and subsequent sealing cover.

Method M11.2

If an electrically insulating cover (of solvent-free epoxy, for example) is applied to the expected anode areas, emission of ions from the reinforcing steel is prevented so that the corrosion process will not take place.

However, the method is only efficient if the reinforcing steel is thoroughly cleaned and the cover is intact (i.e. without any defects). The method should not be applied unless the reinforcement can be completely covered along the perimeter of the reinforcement. Therefore, 100% inspection is a condition for a durable result.

It should be documented that the anchorage capacity of a surface protected reinforcement is identical to that of unprotected reinforcement. If the anchorage capacity is smaller, this should be taken into account.

Epoxy paints (EP) and polyvinyl chloride paints (PVC) are typical insulating products for corrosion protection of reinforcement.

Manufacturer/supplier: control of anodes by corrosion protection

EN 1504-7 specifies requirements for products and systems for application of reinforcement to control anodes with reference to Tables 1–3. According to ASTM A775, annex A1, the following informative requirements for epoxy paint may be assumed:

- The matrix of the paint should be organic, but pigments may be inorganic (e.g. a green colour is preferred in contrast to the red rust).
- The paint should be able to form a film which is free of voids, cracks and other defects.
- Reinforcing bars should be provided with uniform cover over ribs and crests with only slight variation in the film thickness.
- The paint should show chemical resistance to immersion in four aggressive solutions (to be further specified). The paint should not blister, soften, lose bond strength or have other defects.
- The paint should have certain, documented electric resistance to imposed voltage.

- The paint should have a documented, satisfactory resistance to chloride penetration.
- The paint should be able to absorb bending action without crack formation in the tensile side of the reinforcement.
- Reinforcement with paint should have a documented bond capacity when embedded in concrete.
- The paint should have documented long durability against abrasion, impacts and other mechanical handling.

Epoxy paint on reinforcement normally has a thickness of 130–300 mm, applied in a minimum of two coatings. An investigation has shown that optimal balance between cover flexibility and corrosion protection is achieved with a film thickness of 200 mm.

Tar epoxy paint consisting of approximately 40% of coal tar and approximately 60% of epoxy paint is one of the most efficient corrosion protections of reinforcement. Tar epoxy paint is water repellent and very resistant to aggressive substances. However, tar epoxy paint has a high MAL code and should therefore only be used in cases where other epoxy paints and corrosion protection agents have proved to be insufficient.

Epoxy paints on reinforcement can sustain a temperature of approximately 100 °C when subjected to short-term action. When concrete building components reinforced with epoxy-coated reinforcing bars is subjected to fire action, epoxy-painted ribbed bars perform differently from epoxy-painted plain bars, because the ribs resist pull-out of the reinforcement.

In Table E15.1 the properties of corrosion protecting surface treatment, which should be specified, are graded according to methods in EN 1504-7. This means that requirements for a surface protection may depend on the method used (active versus insulating surface protection).

The documentation should be repeated in case of significant changes of a product or in the product specifications:

- When a new mix design is used or when a new type is introduced
- Change of the mix design that may influence the properties of the product
- Change of raw materials which may be significant to the properties of the product.

Table E15.1 Properties or characteristics to be specified for surface protection according to EN 1504-7

Property or characteristics	Test method	Intended use[a] Active protection Method M11.1	Insulating protection Method M11.2
Corrosion protection[b]	?	■	■
Glass transition temperature	EN 12614	□	□
Shear adhesion (painted reinforcement to concrete)[c]	?	□	□

[a] ■, properties and characteristics for all intended uses; □, properties and characteristics for certain intended uses.

[b] The national rules apply where required until an accepted EN test standard for the corrosion protective properties of injection materials is available.

[c] For a given repair project, the designer should decide whether requirements should be made for the anchorage capacity of surface protected steel. So far there is no accepted EN test standard for testing and assessment of the anchorage capacity of corrosion protected reinforcement. Therefore, until further notice, application of DS 2082 'Reinforcing steel. Test of anchorage capacity' is recommended to ensure that corrosion protected reinforcement has the codified anchorage capacity, i.e. $\zeta \geqslant 0.6$ for ribbed steel (DS 411-1999).

Table E15.2   Identification testing of corrosion protection of reinforcement

Properties or characteristics	Test method	Tolerance from declared value
Identification of liquid components		
Colour	Visual	Uniform and similar to the description provided by the manufacturer
Density, pyknometer method	ISO 2811-1	Declared value ±3%
Density, immersed body method	ISO 2811-2	Declared value ±3%
Infrared analysis	EN 1767	Visual comparison
Epoxy equivalent[a]	EN 1877-1	Declared value ±5%
Amine equivalent[a]	EN 1877-2	Declared value ±6%
Volatile and non-volatile matter	EN ISO 3251	Declared value ±5%
Thermogravimetric analysis	EN 1878 and ISO 11358	Visual comparison and loss of mass at 600 °C: Declared value ±5%
Viscosity	EN 1871 and EN ISO 3219	Declared value ±20%
Identification of fresh paint		
Pot life at different temperatures[b]	EN ISO 9514	Declared value ±15%
Consistency	EN 1015-4	Declared value ±15%
Thixotropy	EN 13062	Declared value ±10%
Hardness (shore A or B) after 7 days	ISO 868	Declared value ±3 units

[a] This test is only performed for polymeric repair materials (corrosion protection).
[b] Testing should be performed at different terms, e.g. at 21 °C and the minimum and maximum temperatures (up to 40 °C) recommended by the manufacturer. The test temperature should be kept constant within a tolerance limit of ±1 °C, i.e.: testing should be made according to EN 9514 as a basic standard, but specimens of 1000 ml instead of 300 mm may be used as well as three forms of finishing treatment and test temperatures:

- 21 ± 2 °C.
- Minimum use temperature as specified by manufacturer/supplier ±1 °C.
- Maximum use temperature (up to 40 °C) as specified by manufacturer/supplier ±1 °C.

Note that pot life is defined as the period of time elapsed from when the mixture has been finished and until: for polymeric corrosion protection materials, the mix temperature has risen 15 °C; for hydraulic corrosion protection materials, the filtering stability of the material stops at the value declared by manufacturer/supplier (EN 14497).

Table E15.2 specifies tolerances for identification testing based on the values of properties and characteristics declared by the manufacturer.

Designer/builder: control of anodes by corrosion protection

The EN 1504 standards do not specify requirements for design (and execution) of corrosion protection of reinforcement. As a guide for the choice and design of corrosion protection for repair of corroded reinforcement due to *carbonation*, the following can be listed:

- If the corroding reinforcement has had a covering concrete layer conditioned upon the concrete code, or if the environmental action has been more aggressive or the concrete cover more defective than prescribed in the concrete code, repair should be performed

Table E15.3 Performance requirements for corrosion protection materials for reinforcement

Property or characteristics	Test method	Conformity criteria
Corrosion protection[a]	?	
Glass transition temperature[b]	EN 12614	>10 °K above the maximum use temperature
Shear adhesion[c]	?	

[a]The national rules apply where required, until an accepted EN test standard for the corrosion protective properties of injection materials is available.
[b]10 °K corresponds to 10 °C.
[c]For a given repair project, the designer should decide whether requirements should be made for the anchorage capacity of surface protected steel. So far there is no accepted EN test standard for testing and assessment of the anchorage capacity of corrosion protected reinforcement. Therefore, until further notice, application of DS 2082 'Reinforcing steel. Test of anchorage capacity' is recommended to ensure that corrosion protected reinforcement has the codified anchorage capacity, i.e. $\zeta \geqslant 0.6$ for ribbed steel (DS 411-1999).

with corrosion protection of the reinforcement, for example, in the form of elastic grout with addition of corrosion inhibitor, unless the cover can be sufficiently increased or another form of protection (e.g. surface protection) can be established.

- If a corroding reinforcement has had a covering concrete layer that was not conditioned upon the concrete code (e.g. too thin), it should be investigated whether a conditional cover (of repair mortar) would be sufficient, i.e. without corrosion protection of the reinforcement.
- If it is not possible to establish a conditional cover, the reinforcement should be protected against corrosion either with elastic grout or paint rich in zinc (may be supplemented with surface protection).

The necessary cover thicknesses are specified in concrete codes ENV 1992-1-1 and EN 206-1 or concrete codes valid in current use. It may be assumed that the carbonation depth grows with the square root of time when subjected to constant environmental action.

As a guide for choice and design of corrosion protection by repair of corroded reinforcement due to *chloride attack*, the following can be listed:

- If the corroding reinforcement has had a covering concrete layer conditioned upon the concrete code, or if the environmental action has been more aggressive or the concrete cover more defective than prescribed in the concrete code, repair should be performed with corrosion protection of the reinforcement, e.g. in the form of non-soluble epoxy paint, perhaps supplemented with surface protection of the concrete.
- If a corroding reinforcement has had a covering concrete layer that was not conditioned upon the concrete code (e.g. too thin), it should be investigated whether a conditional cover (of repair mortar) would be sufficient, i.e. without corrosion protection of the reinforcement, perhaps supplemented with surface protection of the concrete.
- If it is not possible to establish a conditional cover, the reinforcement should be protected against corrosion either with epoxy paint, or other measures should be taken, e.g. surface protection of the concrete or electrochemical methods (cathodic protection of the reinforcement).

The necessary cover thicknesses are specified in concrete codes ENV 1992-1-1 and EN 206-1 or concrete codes valid in current use. It may be assumed that the depth of chloride penetration grows with the square root of time when subjected to constant environmental action (Frederiksen *et al.*, 1997).

Contractor/supervisor: control of anodes by corrosion protection

EN 1504-10 on 'Site application of products and systems and quality control of the works', does not specify requirements for execution of corrosion protection of reinforcement.

Therefore, corrosion protection of reinforcement should be applied in accordance with the manufacturer/supplier's instructions.

Corrosion protection

After blast cleaning of the repair concrete, corrosion protection should be applied to the reinforcement if this is specified in the work specification. By application of corrosion protection, the following shall be observed:

- Materials for corrosion protection may be sensitive to reinforcement temperature. Therefore, corrosion protection should not be applied when the reinforcement is subjected to direct sunshine.
- Materials for corrosion protection may be sensitive to strong desiccation. Therefore, corrosion protection should not be applied when the reinforcement is subjected to strong winds unless shielded from the wind.
- Application of corrosion protection should not be made when the reinforcement is subjected to rain unless the repair area is shielded from the rain.
- In case of air-borne chloride on site (e.g. in coastal areas), the corrosion protection should be applied directly after cleaning of concrete and reinforcement.
- The effect of corrosion protection is among other things dependent on whether there is total application or whether there are 'holidays' (especially on the back). Reinforcement with epoxy paint is particularly sensitive to errors where protection is missing or too thin and therefore, such corrosion protection should be checked with a mirror. However, better control is achieved by a pinhole rating meter (e.g. Elcometer).
- By corrosion protection with epoxy paint, a 'sleeve' of alkaline cement grout should first be applied at the borders of the corrosion protection and then, subsequently, the epoxy paint should be applied (to prevent incipient anodes).
- The instructions from the manufacturer with regard to film thickness should be observed. For epoxy paint a film thickness of 150–250 mm (applied in a minimum of two coatings) with 130 mm and 300 mm as the absolute limits should be applied.
- When corrosion protection is applied in several coatings, it is important that the manufacturer's requirements for maximum and minimum interval between operations are observed.
- After application of the corrosion protection, no mechanical preparation of the reinforcement should be performed.

Work environment conditions

When epoxy paint is used for corrosion protection the following conditions should be observed:

- Solvent-containing epoxy paint should be kept away from open fire, since it is combustible.

- The instructions of the national Work Environment Service on works with epoxy and iso-cyanate (polyurethane) should be observed, including:
  - It is required that persons working with epoxy and polyurethane have attended and passed the courses on working with epoxy and isocyanate (polyurethane) offered by the Work Environment Service.
  - Contact with skin and eyes should be avoided.
  - Use of safety glasses and gloves is mandatory.
  - Inhalation of vapour from epoxy paint for long periods should be avoided, for example, by a protective mask.
  - The epoxy paint container should be kept closed.

In case of contact with corrosion protection of epoxy paint, seek first aid immediately:

- Eye contact: Flush with water for a minimum of 10 minutes. Contact a doctor.
- Skin contact: Wash skin immediately with water and soap.
- Inhalation: Go into open air immediately. Contact a doctor.
- Consumption: Do not provoke vomiting. Contact a doctor.

It should be noted that epoxy paint will develop toxic gases when exposed to fire, for example, if reinforcement, which is protected with epoxy paint, is welded.

References

Alekseev, Ivanov, Modry, Schiessel. *Durability of reinforced concrete in aggressive media.* Chapter 9: Special protection of reinforcement. Balkema Publishers, Brookfield, VT, USA, 1993.

Bentur, Diamond, Berke. *Steel corrosion in concrete. Fundamentals and civil engineering practice.* E&FN Spon, London, UK, 1997.

BPS. Typiske beskrivelsesafsnit, betonrenovering. Reparation, injektion og over-fladebeskyttelse. BPS publikation 122. BPS-centret, April, 1997.

CEB. Coating protection for reinforcement. State of the art report. Comite Euro-International du Beton. Thomas Telford Publications, London, UK, 1995.

Frederiksen, Nilsson, Sandberg, Poulsen, Tang, Andersen. HETEK, a system for estimation of chloride ingress into concrete. Theoretical background. Rapport nr. 83. Vejdirektoratet, København, 1997.

Annex E16: Corrosion inhibitors

Introduction

The standard ENV 1504-9 'General principles for the use of products and systems' includes corrosion inhibitors as method M11.3 on 'Use of corrosion inhibitors for repair'. Apart from being briefly mentioned in annex B in ENV 1504-9 under Method 11.3, no instructions for use of corrosion inhibitors for repair exist, not even in EN 1504-3 on 'Structural and non-structural repair'.

In the following, the explanation of method M11.3 is supplemented with knowledge collected from the literature and Danish experience with use of corrosion inhibitor in the repair of reinforced concrete structures.

Background for the method M11.3

A corrosion inhibitor is a substance which will stop corrosion or reduce the corrosion rate when applied in a small quantity. There is a large number of such inhibitors, but few of them are applicable to rehabilitation of reinforced concrete structures (Fontana, 1986). Inhibitors are classified into the following main categories:

- *Anodic inhibitors*, which inhibit corrosion by stabilizing the passive film that has a tendency to break when chloride is present on the surface of the reinforcement. Calcium nitrite is an example of such an inhibitor, which functions by increasing the threshold value in concrete. Addition of calcium nitrite to repair mortar may, however, result in false setting, which should therefore be taken into account (Broomfield, 1997).
- *Cathodic inhibitors*, which are adsorbed on the surface of the reinforcement and act as a barrier to oxidation, which is the cathodic reaction for steel in concrete. However, many cathodic inhibitors such as amines, phosphates, etc., will retard setting of the concrete (Bentur *et al.*, 1997).
- *Organic inhibitors* or adsorption inhibitors act on the anode and the cathode processes, i.e. have double action. Typically, these inhibitors are organic amines (urine), e.g. the amino alcohol type.
- *Vapour phase inhibitors* correspond to adsorption inhibitors, but they act at a high vapour pressure. Therefore, vapour phase inhibitors are applicable for inhibiting atmospheric corrosion without being in direct contact with the metal to be protected.

- *Oxidation inhibitors*. Certain substances, such as chromates, may act as corrosion inhibitors. They are primarily used for inhibiting corrosion of metals and alloys, such as stainless steel.

Anodic inhibitors and organic inhibitors are the most interesting of the corrosion inhibiting substances for reinforced concrete.

It is significant to use a sufficient quantity of corrosion inhibitor for the repair mortar since some inhibitors accelerate corrosion (instead of inhibiting it) in the form of pitting (Fontana, 1986). Therefore, addition of too small a quantity of inhibitor is worse than none at all. On the other hand, care should be taken not to use too large a quantity of inhibitor, which will harm the concrete (Hope and Thompson, 1995). To safeguard against such a situation, the manufacturer/supplier should document that concrete is not harmed by an overdose of corrosion inhibitor. Also, the durability of concrete and the workability of fresh concrete/mortar should not be damaged, see below.

There is a natural limit to the application of certain inhibitors since they are toxic to persons and may accumulate in nature with harmful effects.

Effect of inhibitors

The organic inhibitors may have the following effects:

- *Migration*. If organic (liquid) corrosion inhibitors are applied to untreated concrete surfaces, they will be able to diffuse (i.e. 'migrate') in the concrete towards reinforcing bars (and other metal) to which they affine (Eydemant *et al.*, 1993; Ukrainczyk and Bjegovic, 1993; Phanasgaonkar *et al.*, 1997). Therefore, they are often named 'migrating inhibitors' in everyday language.
- *Threshold value*. If the pore water of concrete contains inhibitors close to a reinforcing bar, the threshold value of chloride in concrete and the critical pH value for onset of corrosion are increased. Laboratory tests have shown that the time of initiation may be increased by approximately 100% (Wiss *et al.*, 1994; Sørensen *et al.*, 1999) when the test methods ASTM G109-92 'Cracked Beam Test' and APM 303, respectively, were used.
- *Corrosion rate*. If concrete contains pore water up to a corroding reinforcing bar, the corrosion rate may be reduced to 15–25% (Masaru, 1998).

The conditions for tests under laboratory conditions and tests performed on site are different. Few site investigations of repair with migrating corrosion inhibitors have been made in practice. However, these investigations have shown that the embedded corrosion inhibitors are active (Sharp-S-658) and that the migration through the concrete in practice may last about a year compared to the rather spontaneous effect that can be observed in laboratory tests (FORCE, 1999).

Manufacturer/supplier: repair with corrosion inhibitors

Requirements for inhibitors

Approximately 25 years ago inhibitors were introduced for application in concrete and repair mortar, etc. During this period no requirements for inhibitors in codes and standards have been made, but Deutsche Institut für Bautechnik is at the present time carrying out investigations of an anodic inhibitor so that this may be the beginning of an impartial analysis of inhibitors. (In Germany, use of repair materials is subject to approval by Deutsche Institut für

Bautechnik.) Based on the requirements made by Deutsche Institut für Bautechnik, the following requirements for (organic, migrating) inhibitors can be formulated:

- They should not pollute the environment, for example, by washing out from the concrete below the ground water table.
- They should be able to increase the threshold value for chloride ions in concrete by at least 100%.
- They should be active in uncracked concrete, as well as concrete with large crack widths, within the limits specified in the concrete codes, e.g. EN 1992-1-1 and EN 206-1.
- They should not change the properties of fresh concrete and repair mortar adversely, neither by overdosing nor by underdosing.
- Local corrosion should be prevented in case of changed concentration of inhibitors, e.g. by washing-out.
- First-year penetration of a migrating inhibitor should be at least 30 mm.

The properties of the inhibitors and their performance should be documented by the manufacturer or supplier using an independent laboratory.

Building owner/designer: repair with corrosion inhibitors

Choice of repair method

The following considerations may be used when choosing a repair method (e.g. with/without inhibitors) (Broomfield, 1997):

- When there is no direct access to an electric power supply, the use of migrating inhibitors should be considered, as many other rehabilitation methods require supply of electric power.
- When there is poor or defective reinforcement continuity (e.g. element construction), use of migrating inhibitors should be considered. Electrochemical rehabilitation methods (i.e. the methods M7.3 and M7.5) will require good reinforcement continuity.
- When the geometry of the building components to be repaired is complicated, use of migrating inhibitors should be considered, as the economy of many other rehabilitation methods depends on the geometry of the building component (electrochemical re-alkalization and chloride extraction as well as cathodic protection).
- In case of building components of prestressed concrete, use of migrating inhibitors should be considered, as certain types of prestressed reinforcement may develop hydrogen brittleness when rehabilitation methods, such as electrochemical re-alkalization, chloride extraction and cathodic protection are used.

There are cases where use of migrating inhibitors need not be considered, because they will not be efficient or they are difficult or costly to use (Broomfield, 1997):

- Cracks in building components formerly injected with polymer (acrylic resin, polyurethane or epoxy) will prevent or inhibit the diffusion of an organic inhibitor (migration) towards the reinforcement.
- Building components with previous surface protection in the form of impermeable paint or similar. Thorough cleaning of the concrete surface may be a rather costly process.
- Building components with previous repair in patches for which dense (polymer-modified or polymer-bound) repair mortars (that may prevent diffusion of the inhibitors).

Choice of inhibitors

The properties of corrosion inhibitors for repair of building components of reinforced concrete should, as a minimum, fulfil the following requirements:

- Be non-toxic to persons on application and not be harmful to nature.
- Be able to migrate in hardened concrete.
- Have long-term effect and be efficient at temperatures that are relevant for the reinforced concrete structures concerned in the relevant environment.
- Inhibit reinforcement corrosion regardless of concentration in the concrete.
- Have minimum effect on the properties of fresh concrete/mortar.

Traditional repair without inhibitors

The traditional form of repair of reinforcing bars which have corroded due to carbonation or chloride attack, can be briefly described as follows:

- *Carbonated concrete*. Remove carbonated concrete around the corroded reinforcing bars. The concrete should be removed in an area of minimum 20 mm behind the reinforcing bar. The concrete along the corroded reinforcing bar should be removed until the reinforcing bar is exposed at least 50 mm into non-carbonated concrete.
- *Chloride-containing concrete*. Remove chloride-containing concrete around the corroded reinforcing bars. The concrete should be removed in an area of minimum 30 mm behind the reinforcing bar. The concrete along the corroded reinforcing bar should be removed until the reinforcing bar is exposed at least 50 mm into concrete where the chloride content is below the threshold value. If the chloride content of the concrete is not measured, the reinforcement should be exposed so that at least 500 mm of the reinforcing bars is not corroded.
- *Repair*. The cleaned reinforcing bars should be given corrosion protection, and removed concrete should be replaced by new concrete (or mortar). Bond improvement between the substrate and the new repair concrete (mortar) may be used.
- *Surface protection*. The repair may be given surface protection to prevent continued carbonation and chloride penetration.

Repair using inhibitors

There are situations where the safety against consequential damages in traditional repairs may be considerably increased:

- *Extensive corrosion*. The entire reinforcement grid of large wall areas in structural members of reinforced concrete may have lost its passivity because it is completely surrounded by carbonated or chloride-containing concrete. This may take place without development of popouts in the concrete due to corrosion. It is not possible to remove the concrete behind the reinforcement grid everywhere on the large wall surfaces without setting up a temporary, comprehensive bracing of the structural member. Use of migrating corrosion inhibitors may remedy this situation.
- *Incipient anodes*. When traditional repair of reinforced concrete with corrosion damage is made, there is a risk that incipient anodes will develop around the repair. The original anodes will be passivated by the repair and the ambient concrete may therefore create anodes on through-going reinforcing bars due to differences of pH value, chloride content or defects

between the repair concrete and the original concrete. It has proved to be possible to limit the probability of incipient anodes by applying migrating corrosion inhibitors in the specified quantity (ml/m^2 concrete surface) over an area of at least 500 mm around the repair.

For use of organic inhibitors by patch repair, the procedure is, in principle, as follows:

- Remove poor and damaged concrete around corroded reinforcing bars.
- Clean corroded reinforcing bars by steel brushing to at least St 2 (also the back).
- If corroded concrete has been cut off mechanically, the cutting surface should be cleaned so that microcracks are removed. If, however, the concrete has been removed by waterjet cutting, cleaning of the concrete will normally not be necessary.
- Protect the reinforcement by cement-based corrosion protection with addition of organic inhibitor.
- Spray the specified quantity of migrating inhibitor on to the cutting surface of the concrete and the ambient concrete surface.
- If necessary, the cleaned cutting surface is provided with cement-based bond improvement. This bond improvement may be supplemented with addition of a migrating inhibitor.
- Fill the repair hollow with repair mortar with addition of an organic inhibitor.
- Provide the cleaned member and the repair with a surface protection to prevent further carbonation and chloride penetration.

As the corrosion protection of the reinforcement, the bond improvement of the concrete and the repair mortar contain organic, migrating inhibitor, these inhibitors may migrate through the concrete and protect reinforcement which was not given sufficient protection from incipient anodes by the repair.

Contractor/supervisor: repair with corrosion inhibitors

Repair with corrosion inhibitor should fulfil certain minimum requirements to comply with the 'spirit' of EN 1504-10 on 'Site application of products and systems and quality control of the works'. Documentation of the corrosion inhibitors used, the workmanship and the finishing treatment should be available.

Conditions

Repair of building components of reinforced concrete should be performed according to project material which normally comprises work specification and drawings, etc., fulfilling the requirements in the EN 1504-10 on 'Site application of products and quality control'.

It is a precondition for execution of the work that a registration of state and an assessment of the concrete and reinforcement of the building component are available together with the designer's evaluation whether it is possible to repair the building component by controlling the corrosion state of the reinforcement by inhibitors. Furthermore, the project should comprise verification of the load-carrying capacity of the building components taking due account of corrosion damage on the reinforcement. If the load-carrying capacity of the building component is insufficient according to codified requirements, rehabilitation shall – apart from repair of the reinforcement using corrosion inhibitors – also include strengthening of the building component according to principle P4 (i.e. 'Strengthening of building components') in ENV 1504-9 on 'General principles for application of products and systems'.

The concrete

It is assumed that the concrete is checked for cracks and defects and that the project specifies how to repair these defects in connection with repair of the corroded reinforcement using inhibitors.

The defects, cracks, etc., in the concrete should be repaired in accordance with principle P3 (i.e. 'Replacement of damaged concrete') in the standard ENV 1504-9 on 'General principles for application of products and systems'.

The reinforcement

The corrosion state and carbonation depth of the reinforcement are significant for the extent of the repair.

State of corrosion

The state of corrosion of the reinforcement will normally be determined in connection with inspection and preliminary testing of the building component concerned. Therefore, determination of the corrosion state of the reinforcement is not necessary unless required in the project work specification.

If corrosion has been detected in the reinforcement of the building component resulting in loss of cross-sectional area of the reinforcement, the need for strengthening, if any, should be documented before the corroded reinforcement is repaired, see principle P4 on 'Strengthening of building components' in ENV 1504-9 on 'General principles for application of products and systems'.

Carbonation of the concrete of the building component

The carbonation depth of the concrete will normally be determined in connection with inspection and preliminary testing of the building component. Therefore, determination of the carbonation depth of the concrete is not necessary unless required in the project work specification.

The carbonation depth in the building component is significant for the quantity of concrete to be removed and the choice of carbonation retarding surface protection of the building component.

Chloride penetration into the concrete of the building component

Normally, the chloride content of the concrete will be determined in connection with inspection and preliminary testing of the building component. Therefore, determination of the chloride content and the penetration depth is not necessary unless required in the project work specification.

The depth of chloride penetration into the building component is significant for the quantity of concrete to be removed and the choice of carbonation retarding surface protection of the building component.

Reception control

The standard EN 1504-10 on 'Site application of products and quality control' does not specify special conditions for reception control of corrosion inhibitors. Based on the literature and Danish experience, the following requirements are specified:

Reception control of corrosion inhibitors

In the work specification, the inhibitor will (in principle) be specified by type and conformity criteria, either as performance requirements or as numerical conformity criteria for measurable properties and characteristics of the materials with corresponding test methods and criteria of acceptance. Based on these requirements, the contractor will normally choose the inhibitor.

On reception of the inhibitor on site it should be established that the inhibitor corresponds to the expectations (as ordered). This may be done visually by the following identification testing for inhibitor:

- Label
  - *Test method*. Visual inspection.
  - *Requirements*. Identical with the product ordered.
- Colour
  - *Test method*. Visual inspection.
  - *Requirements*. Uniform and identical with the manufacturer's specification or a colour test.
- Corrosion
  - *Test method*. Two small containers are filled with seawater or a 3% kitchen salt solution. Furthermore, 0.5% inhibitor is added to one container. A clean piece of reinforcement steel is placed in each container.
  - *Requirements*. No initiation of corrosion on reinforcement in the container with inhibitor.

Reception control of repair mortar and other repair materials

Reception control of repair mortar and other repair materials is performed as specified in the standard ENV 1504-9 for method M3.1 on 'Hand application of repair mortar', method M3.2 on 'Recasting with repair concrete' and method M3.3 on 'Application of sprayed mortar or sprayed concrete'.

Control and documentation of the works

In planning of the control of the works it should be taken into consideration that corrosion inhibitors are normally colourless.

Control of the applied quantity of inhibitor

The concrete surface should be dry when the inhibitor is applied. The inhibitor is sprayed or rolled on to the substrate in full saturation. It should be documented that the applied quantity of inhibitor is at least as specified in the project work specification for the individual areas of the concrete surface. The procedure according to the principle in the standard ISO 7254-1986 on 'Paints and varnishes – Assessment of natural spreading rate – brush application' should be followed.

The inhibitor consumption should be registered either as kg or litres of inhibitor, e.g. corresponding to the area of a concrete surface (m^2). The increase of the surface area due to a roughness, e.g. the cut concrete surface in the repair hollow, should be estimated beforehand, e.g. during the test repair. Further, it may be necessary to know the density of the inhibitor. The inhibitor is assumed to be uniformly applied. The calculated intensity of the inhibitor applied should be given in the unit ml/m^2 concrete surface or g/m^2 concrete surface.

Control of applied bond improvement, corrosion and repair mortar

Control of bond improvement applied to concrete, corrosion protection of reinforcement and applied repair mortar should be performed as specified in the standard ENV 1504-9 for method M3.1 on 'Hand application of repair mortar', method M3.2 on 'Recasting with repair concrete' and method M3.3 on 'Application of sprayed mortar or sprayed concrete'.

Safety and health

ENV 1504-9 on 'General principles for application of products and systems' does not specify requirements for safety and health apart from stating that the national rules should be obeyed.

Therefore, the repair materials used should fulfil the requirements made in the work specification and specified by the national Working Envionment Service, see Instruction no. 1017 of the Danish Ministry of Labour on 'Arrangement of building sites and similar places of work', dated 15 December 1993. This entails, among other things, that repair materials should be available on site in their lawful, labelled original packaging.

References

Bennett, Bartholomew, Bushman, Clear, Kamp, Swiat. Cathodic protection of concrete bridges: a manual of practice. Strategic Highway Research Program, SHARP-S-372. National Research Council, Washington, DC, USA, 1993.

Bentur, Diamond, Berke. *Steel corrosion in concrete, fundamental and civil engineering practice*. E&FN Spon, London, England, 1997.

Brameshuber, Brockmann, Rössler. Textile reinforced concrete for formwork elements – Investigations of structural behaviour. Procedings of the Fifth International Conference on Fibre-Reinforced Plastics for Reinforced Concrete Structures. Cambridge, UK, 2001.

Broomfield. *Corrosion of steel in concrete, understanding, investigation and repair*. E&FN Spon, London, England, 1997.

Eydemant, Ostrovski, Demidov. Analysis of diffusion rate of migrating corrosion inhibitor MCI 2000 in concrete using radioactive isotope tagging techniques. Test Report. Institute of Construction Materials and Kurchatov Institute of Nuclear Physics, Moscow, Russia, 1993.

Fontana. *Corrosion engineering*, Third Edition. McGraw-Hill, New York, USA, 1986.

Hope, Thompson. Damage to concrete induced by calcium nitrite. *ACI Materials Journal*, September–October 1995.

Højlund Rasmussen. Betonindstøbte, tværbelastede boltes og dornes bæreevne. Bygningsstatiske Meddelelser, årgang XXXIV, 1963. Teknisk forlag København, 1963.

Ibrahim, Ferrier, Curtil, Hamelin. The role of composite materials in concrete durability. Case study of alkali-aggregate reaction. Procedings of the Fifth International Conference on Fibre-Reinforced Plastics for Reinforced Concrete Structures. Cambridge, UK, 2001.

Jensen, Petersen, Poulsen, Ottosen, Thorsen. On the anchorage to concrete of SikaCarbo-Dur CFRP strips, free anchorage, anchorage devices and test results. International congress on creating with concrete. Dundee, Scotland, 1999.

Kessler, Powers, Lasa. Update on sacrificial anode cathodic protection of steel reinforced concrete in seawater. Corrosion 95, Paper 516. NACE International, Houston, TX, USA, 1995.

Lacasse, Labossière, Neale. FRPs for the rehabilitation of concrete beams exhibiting alkali-aggregate reactions. Procedings of the Fifth International Conference on Fibre-Reinforced Plastics for Reinforced Concrete Structures. Cambridge, UK, 2001.

Lynch, Duckett. Bible Christian Bridge, Cornwall – Advance composite material wrapping trial.

Highway Agency Maintenance Conference. Nottingham University. Nottingham, UK, 1998.

Masaru. MCI-2020 Long Term Test. Protection of rebar in concrete. Interim Report. General Building Research Corporation of Japan, Osaka-Fu, Japan, 1998.

Mays, Hutchinson. *Adhesives in civil engineering*. Cambridge University Press, Cambridge, UK, 1992.

Phanasgaonkar, Cherry, Forsyth. Organic corrosion inhibitors; How do they inhibit and can they really migrate through concrete? Australasian Corrosion Conference Brisbane, AMECR 97/30. Queensland, Australia, December 1997.

Sørensen, Poulsen, Risberg. On the introduction of migrating corrosion inhibitors in Denmark – A review of documentation tests and applications. Proceedings of International Conference on Infrastructure Regeneration and Rehabilitation. University of Sheffield, England, 1999.

Ukrainczyk, Bjegovic. Diffusion of the MCI 2020 and MCI 2000 Corrosion inhibitors into concrete. University of Zagreb, Croatia, 1993.

Wiss, Janney, Elstner Associates Inc. Cracked beam corrosion tests. Test Report WJE no. 922041. Cortec Corporation, St. Paul, MN, USA, 1994.

Annex F: Quality control

Introduction

The standard EN 1504-8 on 'Quality control and evaluation of conformity' specifies requirements for quality control of repair materials and systems and for evaluation of conformity with the requirements, and for design of labels for marking of repair materials and systems for rehabilitation of concrete structures according to EN 1504-2 to EN 1504-7. EN 1504-8 lists the general principles, but assumes that the reader is familiar with control of repair materials and systems.

On control

In the following a brief explanation of the fundamental principles of quality control based on EN 1504-8 is given.

Requirements in EN 1504-8

In EN 1504-8 it is required that fulfilment of requirements for repair materials and systems in EN 1504-2 to EN 1504-7 is verified by the following forms of control:

- *Identification testing*, i.e. testing of properties in cases where a repair material or a system is required to conform to certain requirements. Identification testing is made prior to the use and is meant for identification of the repair material by testing simple properties by one or more of the CEN test methods.
- *Performance testing*, i.e. testing of properties in cases where repair material or a system is required to conform to certain requirements at any time during application and use. Typically, performance testing is made to determine a property, e.g. to be used in the design.
- *Factory control*. The manufacturer should perform FPC (factory production control) to ensure continuous conformity of protection with the requirements made for the repair material or system according to the relevant EN 1504 standard, e.g. according to EN ISO 9001. Thus, FPC can be based on own control to ensure that all supplies of the repair material or system concerned fulfils the requirements made in the relevant EN 1504 standard. According to EN 1504-8, FPC is required to include:
  - Inspection, sampling and testing of raw materials, production equipment and work processes
  - Inspection, sampling and testing of the finished repair material or system.

Observations from the testing should be treated by relevant, statistic methods.

Sampling

Sampling forms the basis of FPC. Raw materials, as well as finished repair materials and systems, should be divided into control sections, and samples are drawn from these control sections. For a continuous production of a repair material, it is convenient to define a control section delimited by mass, volume or time. Similar conditions apply to raw materials.

Sampling from a control section should be performed so that all samples are evenly distributed in the control section and are representative of the control section. Each sample should be unambiguously marked with sampling location and time of sampling. The sample size should be sufficient for testing according to the required test method. Part of the sample should be kept as reference. The necessary information of the test should be registered, including:

- Date of manufacture and sampling
- Unambiguous identification of the repair material and the drawn sample
- Type of repair material and quantity of the sample
- Name of manufacturer
- Manufacturer's batch number/charge number from which the sample was drawn
- The quantity of the manufacture, which is represented by the drawn sample
- Physical condition of the sample
- Colour or appearance of the sample
- Name of the person, who has drawn the sample
- Sampling method.

Measures in case of changes of manufacturing conditions, raw materials, composition, etc

If the raw materials, the composition or the conditions of manufacture of a repair material or system are changed, and if these changes result in changes of the documented properties, identification testing and performance testing according to the relevant EN 1504 standard should be performed.

Measures in case of non-fulfilment of requirements

By FPC the manufacturer should ensure that all materials, inclusive of raw materials, packing and finished repair materials or systems, which do not comply with the requirements made in the relevant EN 1504 standard, are clearly marked and separated from the rest so that their use and despatch are prevented.

Marking and label

When repair materials are marketed in containers (e.g. buckets or bags) they should be clearly marked with the following information. For bulk materials and systems, similar information should be supplied. The information comprises:

- Name and address of the manufacturer, the registered product/trade name and other identification of the manufacturer or agent authorized by the manufacturer established within the EEA countries (European Economic Area), i.e. the EU and EFTA countries, as well as information on place of manufacture.

- Identification of the repair material or systems, i.e. registered trade mark and batch no.
- Type of repair material or system with no. and date of the relevant EN 1504 standard.
- A summary of storage conditions comprising among others the durability with clear indication saying: 'This repair material does not necessarily fulfil requirements in EN 1504 after ... (date)'.
- Instructions for use including warning against use under special conditions and special national restrictions.
- For CE marking, reference is made to annex ZA in the relevant EN 1504 standards, see annex D.

Index

Printed and bound by CPI Group (UK) Ltd, Croydon, CR0 4YY

22/10/2024

01777459-0001